Praise

Contemporary Exprience

David Moffett-Moore provides us with a succinct and inspiring synthesis of contemporary creation theology, informed by modern insights from both science and scriptural research. However, this is more than an inspiring read: it provides a spiritual and ethical blueprint for how we live our faith in the 21st century. In the author's own words: "If we see all creation as God's gift and God's property, even an expression of God's self-revelation, how do we demonstrate that belief through our behavior?" (p. 47)

Diarmuid O'Murchu
Author of *Quantum Theology* and *God in the Midst of Change*

I am delighted that Energion's series of books on Creation now includes a volume on *Creation in Contemporary Experience*. Moffett-Moore succeeds in engaging the Christian tradition about Creation in a meaningful dialogue with our contemporary understandings of the natural world. He insists that all understandings of God are only partial, but, if they are big enough, believers who hold them can confidently and profitably benefit from the scientific discoveries that inform the way in which we live today. A God that is "expansive and inclusive, inviting of opposites and welcoming of options" is the God that is constantly revealed to contemporary experience. Appealing to Thomas Aquinas' demonstration that a proper understanding of Creation must be the basis for an understanding of the Creator, this book challenges anyone who claims to have grasped God's being and will with its central question: Do not all the changes in our scientific understanding of the universe and the world in which we live demand changes in our understanding of God? Answering this question Moffett-Moore shows the inadequacy of our contemporary public atheists and attempts to offer a twenty first century understanding of God. Readers of this book will gain new insights into the mystery of the Creator of Creation

Herold Weiss, PhD
author of *Creation in Scripture* and Professor Emeritus of New Testament, St. Mary's College, Notre Dame

Although there are partisans on "both sides" of the issue, for many of us there is no war between science and faith. We can affirm the premise that God is the Creator without having to abandon accepted scientific theories, including evolution. In *Creation in Contemporary Experience* David Moffett-Moore offers a thoughtful pastoral testimony to the creative presence of God in the evolutionary process. He doesn't pretend that God fills the unfilled gaps, but he does invite us to discern the presence of God in the world around us, allowing us to reclaim the word Creator from those who would reject modern science. And for that we should be thankful.
Robert D. Cornwall, Ph.D.
Author of *Worshiping with Charles Darwin*

Creation in Contemporary Experience introduces us to the science of Creation without challenging the Biblical story of it, thus enabling the reader to develop a theologically comfortable perspective on it.
Rev. Dr. Douglas E. Busby, MD, MSc, DMin

Creation in Contemporary Experience

David Moffett-Moore

Energion Publications
Gonzalez, FL
2014

Copyright © 2014, David Moffett-Moore

Cover Design: Henry Neufeld
Cover Image: © Jorisvo I Dreamstime.com

ISBN10: 1-63199-010-1
ISBN13: 978-1-63199-010-6
Library of Congress Control Number: 2014938153

Energion Publications
P. O. Box 841
Gonzalez, FL 32560

energion.com
pubs@energion.com
850-525-3916

Dedicated to my father, the Rev. John E. Moore:
Eagle Scout, avid gardener and amateur biologist,
lover of nature and of nature's God,
who taught me that thinking and believing
are not mutually exclusive

Acknowledgements

I want to begin by thanking Henry Neufeld and Energion Publications for undertaking this series on Creation. It is a risk of faith. Energion calls itself a publisher for the creative Christian mind, and this series offers ample proof of that statement. It has also given me opportunity to express my interest in contemporary scientific understandings of the world within which we live and experience the divine. I intentionally say "interest," not "understanding." I am an amateur and layman in science, whether quantum physics or chaos theory and evolutionary biology, but I have great interest. So thank you Henry and Energion for giving me this opportunity to express that interest.

I also thank my fellow authors in this series: Harold Weiss, author of *Creation in Scripture*, Edward Vick, author of *Creation: The Christian Doctrine*, Robert Cornwall, author of *Worshiping with Charles Darwin*, and Tony Mitchell's upcoming *Creation: The Science*. I hope my participation in this series is up to the high standard they have set.

I thank my good friend Rev. Dr. Doug Busby for his time and effort in proof reading and improving my rough manuscript. Doug holds Doctor of Medicine, Doctor of Ministry, Master of Divinity and Master of Science in bio-physics, has worked for the Cleveland Clinic, Continental Airlines, and N.A.S.A and written on space medicine. His current interest is in the role of spirituality in health and healing. Every time I talk with him, I learn something.

I owe a huge debt of gratitude to Chris Eyre, whom I have never met. Chris is a solicitor in England (that would be an attorney in the states) with a degree in physics who serves as a copy editor for Energion and for this manuscript. I deeply appreciate Chris's scientific knowledge and objective eye that has helped so much to improve my text.

Thanks to the congregations I have served, St. Peter's United Church of Christ in Frankfort, Illinois and now Portage United Church of Christ in Portage, Michigan, for allowing me time to read, to write and most of all to ponder.

Thanks to my beloved wife, Becki, for her support, encouragement and patience, for listening and saying "That's interesting."

I have very much enjoyed all the reading, thinking and most of the writing. I confess that, despite all the help from others, I am sure I have made some mistakes. I admit they are my own! I crave the reader's forbearance and hope that my errors, whether in grammar or of scientific fact, do not deter from your reading the book or reflecting on its ideas. Part of the reason for my existence on this earth is to prove that God is gracious!

Table of Contents

Acknowledgements v
Foreword..ix
All the Creation Accounts of the Bible 1
From Creation to Creator..7
A Contemporary Creation Story...11
Evolutionary Biology and the Eighth Day 17
Toward a Theology of Evolution...21
Creation and the Divine Feminine..29
Quantum Physics and the Dance of the Cosmos.....................33
Chaos Theory and the Risk of Freedom 41
Caring for Creation: On a Christian Ethic.................................47
Christ and the New Cosmology 55
A Contemporary Expression of the Eternal God......................59
Postscript 65
Bibliography 67

Foreword

Before beginning this book, the reader needs to know my approach to the reason and faith, science and religion debate and my approach to the Scriptures.

I grew up in a Methodist parsonage: services every Sunday morning and evening, spring revivals, singing in the children's choir, active in the youth group, Vacation Bible School and church camp every summer and of course in worship every week. I grew up on the inside of the faith, knowing I was a Christian even before I was able to choose being one. I can remember times in my life that I have not been a good Christian, but there has never been a time I did not consider myself Christian.

I also grew up in a family that had both Charles Darwin's *Origin of Species* and his *Descent of Man* on our bookshelves, besides books identifying birds and trees and exploring the geography and geology of the planet we inhabit and three sets of encyclopedias. Mom grew up on a farm, playing in the woods and was better than most of the boys in most sports. Dad grew up outdoors as much as indoors, achieved the Eagle rank in Scouting, and studied biology for a time in college. When we went on a family vacation, it was always tent camping and hiking in the woods.

For me, being a Christian and accepting evolution has never been contradictory.

Mr. Bishop taught biology every year in high school and alternated between chemistry and physics. As a chemistry student, he had combined liquids of his choosing rather than the teacher's and blew the lab up. They hired him as a teacher in spite of his reputation as a student. He took us outside for those explosions. He liked being "explosive."

As a freshman in Mr. Bishop's biology class, he suggested that I write my term paper on what the Bible said about creation and evolution. Being young and naïve and trained to please my elders,

I quickly agreed. He thought that he was setting me up, knowing I was "the preacher's kid." I went home and asked Dad what the Bible said about creation and evolution. His response was totally unexpected by both me and Mr. Bishop.

"Which place?" Dad went on to explain that there were perhaps a dozen different passages of scripture that described God's work of creation, each of them different and unique; no two passages telling the same story in the same way. I was immediately confused.

We talked about creation and evolution, science and faith, reason and religion. He kept coming back to the same point: they are complementary, not competing approaches. They each ask different questions and therefore it should be expected that they have different answers. That does not mean that the answers contradict each other, only that they are different, which we knew from the beginning.

We lived in Dunkirk, Indiana, which was the glass capital of the state; more glass was made in Dunkirk than anywhere else in Indiana. Dad took a pop bottle. "There are two ways of looking at this bottle. You can study its beauty and symmetry, how useful it might be, how well it fits the hand and appeals to the eye. Or you can flip it over and look at the mold mark on the bottom of the bottle. That will tell you where it was made and when it was made, even within a few weeks on the calendar." I knew this was true because in Dunkirk at that time, we always looked at the mold mark before opening the bottle.

"These are different ways to look at the same bottle. It's not that one is right and the other wrong, they are simply different: different perspectives, different questions, different approaches and, of course, different answers. And this is just a bottle! Imagine how different we can be in approaching the creation of the universe!"

Mr. Bishop received a term paper that he did not expect, but instead one that described the different creation accounts in scripture and the different approaches to creation from scripture and from science.

I approach science and religion as mutually compatible, complementary disciplines. Each has its own premises, perspectives and approach; it is not a question that one is right and the other wrong, they are simply different. Albert Einstein, perhaps the greatest scientific mind of the twentieth century, said, "Religion without science is lame; science without religion is blind." I won't argue with him; we need both. My approach in this book is that science and religion are friends, not enemies. They can and should work together. We will see how they do.

Part of the tension between science and faith is in how we approach the Bible. Is it the inerrant word of the one true God, dictated to human scribes in both writing and translating without their participation in the process, or is it the inspired writing of perhaps hundreds of authors, working collectively with the divine spirit in a great and holy creative process? The house I grew up in, again, a Methodist parsonage and Christian home, affirmed the latter.

The Bible includes writings accumulated over fifteen hundred years or more, describing events from the beginning of time to the end. It includes many different styles of writing and forms of literature. Dozens of authors, many anonymous, working with the inspiration of the Holy Spirit, recorded their experiences and understandings of the God who created them to be partakers of the divine nature, participants in God's glory and co-creators with God. Such a God, who made us out of love and for love to be God's children and partners, would not then simply turn off the minds of the writers and use them as we might use a pen to write our story. The writers of the Holy Scriptures were themselves holy and engaged in a holy activity. The process of the Bible, from its writing through its compiling to its translating is an inspired, holy activity, all of it under the direction of the divine.

I take the Bible far too seriously ever to simply take it literally, or literalisticaly. It is not a simple book, for a simplistic understanding does not honor either the book or the divine presence revealed through it. I study the Bible; I worship God. To worship the Bible would be idolatrous.

The Letter to the Hebrews states, "The word of God is living and active, sharper than any two edged sword, penetrating us and separating soul and spirit as joints and marrow. It discerns our thoughts and attitudes. Nothing in all creation escapes; everything is laid bare before the one to whom we must all give account" (Heb. 4:12-13). Paul writes to Timothy, "All scripture is inspired by God and useful for teaching, discipline, correcting and training in righteousness, so that we may be equipped and prepared for every good work" (2 Tim. 3:16-17). Every time I study scripture, it studies me. In my experience, the Bible is a living entity, studying and exposing me, opening me up and revealing truths that I would deny about myself and opening me to a God who so greatly loves, accepts and welcomes me. Bible study is a partnership between the Holy Spirit and me to reveal to me the eternal Word of the living God. I will not simplify or denigrate this sacred process and divine partnership.

What follows in the text of this volume will be a perspective that holds faith and reason, and science and religion, as partners in an approach to Scripture that studies it intelligently and seeks eagerly the divine presence that is within it and is revealed through it. We will explore Creation itself as a revealing of the divine presence, as sacred as any ink printed on any paper, and we will affirm our place and purpose in the divine process of God's Creation.

All the Creation Accounts of the Bible

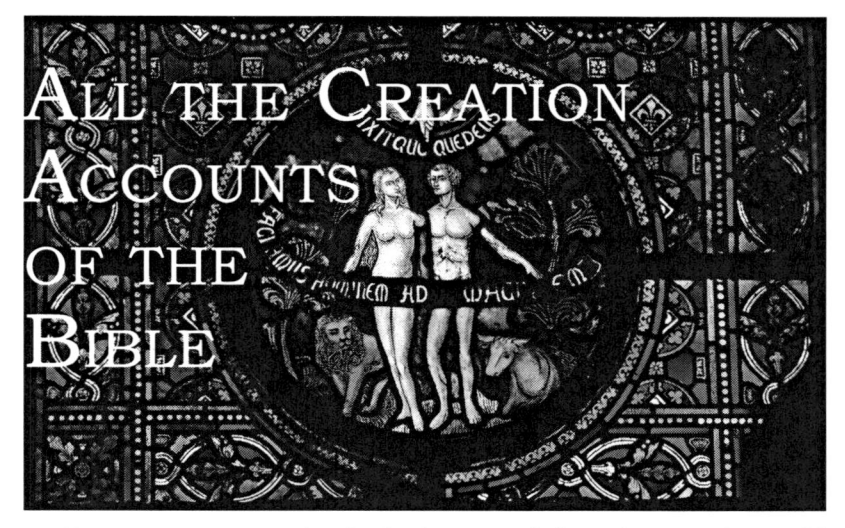

As a teen I was shocked when my father, the preacher, told me there were as many as a dozen different Creation accounts in scripture. How this could be, I wondered. Since there is only one creation, how can there be so many different accounts of it? Wasn't the Creation story that begins the Bible, the one I had learned as a child, the true one? My father explained that each Creation account had its own purpose and should be accepted on its own terms. In other words, let the Bible be the Bible: what it is rather than what I want it to be.

Genesis chapter one begins the Bible with those striking, powerful and appropriate words, "In the beginning . . . "Genesis 1:1 - 2:3 offers an account as beautiful as it is familiar. God is very well organized! On day one God creates light. On day two God creates waters above and below. On day three God creates the dry ground and vegetation. On day four God creates the individual lights. On day five God creates everything that swims in the sea or flies in the sky. On day six God creates the animals of the field and the plants for food. Finally, as the last act of creation, God creates human beings in the image and likeness of God. All this simply by speaking, "let there be!" and there is.

We studied Genesis 1:27 in my Introduction to the Old Testament class in seminary. Dr. Charles Kraft explained that Hebrew poetry is different from English poetry. English poetry typically has similar sounds; Hebrew poetry has similar ideas. Genesis 1:27 says:

> "God created humans in God's own image;
> In the image of God, God created them:
> Male and female God created them."

In using this structure, the author of the passage is telling us that the image and likeness of God is bound into our sexuality. Human sexuality is a sacred gift and can be seen as a revelation of the divine within us. This is a powerful message in a culture that sees human sexuality as anything but sacred!

Another often-quoted verse immediately follows, "God blessed them and said to them, 'Be fruitful and multiply. Fill the earth and subdue it. Rule over every living creature" (Gen. 1:28). This verse has been used to justify our re-engineering of the planet: strip mining, deforestation, pollution, etc. based on the attitude that God gave the planet to us and said we could do whatever we want with it. But that is not strictly what the passage says. The Hebrew words translated as subdue and rule also carry the sense of governing and managing, at least implying an obligation for responsible use of available resources.

When James Watt was President Reagan's Secretary of the Interior, he advocated the maximum use of all our natural resources. He based his political position on his faith. As a Christian, he believed Christ would return any day, that we were truly in the End Time, and so we could exhaust our resources and pollute our planet without regard for consequences because Christ was coming soon and very soon and we would be saved from our actions. The only thing wrong with this theory is that it proved wrong. So far, every prediction of the end of time has proven false. My opinion on how we treat our earthly home is that we should not trash it until we find and inhabit another planet. We may sincerely believe that we are living in the End Times, but I suggest we still act cognizant of the possibility that, like those who have anticipated it through the centuries, we too might be wrong.

One evening in seminary a guest speaker gave a lecture on the first chapter of Genesis and the history of the universe as science has hypothesized it to be. The lecturer was an astrophysicist from Northwestern University. With slides of pictures, charts and graphs he explained that, at least as a general overview, the Big Bang Theory and the Genesis chapter one story are reasonably compatible, with a similar order of events. I sat there thinking it seemed a bit of a stretch, and unnecessary, as the Big Bang Theory and Genesis chapter one had different perspectives and were generating different questions to which we should reasonably expect different answers.

One example was that in Genesis God creates light on the first day and yet does not create the sun, moon and stars until the fourth day. How can this be, we might ask. How can there be light without some object to give the light? Yet that is precisely what happened in the Big Bang Theory: the unimaginably powerful explosion that begins Creation radiated immense heat and light, yet the stars did not coalesce for hundreds of thousands of years. Both in Genesis and the Big Bang Theory we have light before there are objects to radiate light.

Creation Accounts in Genesis

In Genesis 1:1-2:3 we have the most familiar of all the Bible's Creation accounts. Genesis 2:4 immediately begins a different account, saying clearly "This is the account of the heavens and the earth when they were created. When the Lord God made the earth and the heavens." In Genesis 2:4-25 we have a totally different and probably older account of Creation. It has its own purpose and order. There is a tendency to graft the two different accounts into one combined account, which has its own difficulties. We can't combine these two different accounts without losing something in the process. Genesis One begins with the creation of the material universe and ends with the creation of human beings. Genesis Two assumes the universe and begins with the creation of humans, focusing more on our divine-human relationship.

In this second account we have a God that is much more involved in the work of creation. In Genesis 1, God simply speaks and there is; God is majestically detached from the work of creation. In Genesis 2, God makes and forms and gets God's hands dirty, and even plants a garden. God breathes into the dust caked on the palm of God's hand to create a human being. We get names for the first humans. "Adam" is transliterated rather than translated and the Hebrew word for human becomes a masculine name. "Adam" is to "Admah" what "human" is to "humus." These words all mean of the earth, of the dirt and clay that was in God's hand, and could just as accurately be translated "earthling." That is what we are: of the earth and alive only by the breath of God. "Eve," which literally means "living being" likewise becomes a transliterated feminine name. The fact that these words are so closely interwoven in the account should tell us that we are not dealing with a scientific analysis, but with words that are more like poetry. The fact that in Genesis 1 the last act of Creation is the creation of human beings and in Genesis 2 the first act of Creation is the formation of human beings should be proof enough that we cannot take the Bible literally if we are to take it seriously. This is not just a numbers game.

Acting as a balance to the sometimes rapacious interpretation of Genesis 1:28, in 2:15 we have God placing the earthling in the garden that God planted in 2:8, with instruction to take care of it, to till the garden of God and tend the flock of God. All Creation still belongs to God; it is in our safekeeping and we will be held accountable for our care of it.

Other Old Testament Accounts

The next Biblical account of creation we have is in God's interrogation of Job that begins in chapter 38 of the Book of Job. God has had enough of Job's railing against the heavens and repeatedly questioning why, as a righteous person, he has experienced misfortunes. Instead of answering any of Job's questions, God responds with questions of God's own, "Where were you when I laid out the foundations of the earth, when I set the limits and measured the dimensions? Who is this that darkens my counsel with words without understanding? I will question you and you answer me" (Job 38:1-5, arranged)! There is no question of God's power or right. God is God and that is that! Job responds to this divine revealing the only way he can, with words every mystic craves to confess, "Before, I had only heard of you with my ears. Now with my own eyes I have seen you. I submit myself and repent in dust and ashes" (Job 42:5-6).

The heart of faith is never creed or doctrine, but an incredible and ineffable divine-human encounter, experiencing the reality that is God for one's self, that can never be clearly understood and rarely be fully described, yet can never be denied. Not hearing or reading or talking about God, but a direct divine-human encounter is the beginning and foundation of faith.

The Psalms are filled with descriptions of creation: Psalm 8, 19:1-6, 24:2, 33:6-9, 102:25-28, 104, 147 and 148. One could pick out additional verses scattered throughout the Psalms that describe God as Creator, that praise God for Creation and recognize God's rule over it.

Proverbs 8 recounts all creation in the creation of holy wisdom, identifying wisdom as the master craftsman, the architect, the delight of the God who by wisdom laid out Creation. In both the Hebrew and Greek languages, the wisdom of God is always in the feminine. Ecclesiastes 3;1-15 recognizes God as Creator, Creation as good and beautiful, and admits that we cannot understand all that God has done. Isaiah 40:21-31 recounts God's act of Creation and God's acts through human history.

New Testament Accounts

The New Testament offers its own collection of Creation accounts, though not as many as the Old. The prologue in John's Gospel, 1:1-18 begins as Genesis does, "in the beginning" and declares the sacredness and purposefulness of God's Creation. Romans 1:18-20 states that God has made God's presence and plan plain before all people since Creation. Colossians 1:15-20 describes Christ's role in Creation and rule over it. In John and Colossians, Christ plays the role of divine wisdom as in Proverbs 8. The book of Revelation is traditionally confined to the End Time rather than its beginning, yet it describes God as Creator, describes Creation as belonging to God and declares that God is not finished with Creation. I like to say that on the seventh day God rested, but it nowhere says God quit creating! Especially in Revelation, God comes to us from the future as much as from above. Indeed, Heaven may be a factor of time rather than of place.

Here we have over a dozen scripture passages that describe the divine act of Creation. Some are long, some short. Each is different. Yet they also share a strong central theme, which can be seen as three strands of a common cord. First, God is Creator. Second, Creation is both inherently good and a revelation of the divine presence. Third, we have a place and a purpose in God's Creation. This is the point of the Bible's Creation accounts, all of them. It answers the "why" question of religion and cannot be found by any of the "how" questions of science.

From Creation to Creator

The question of Creation is essentially a question of our own existence. Every four year old asks, "Where did I come from?" At four they don't need a detailed analysis of the biological process. When my sons were four to six years old, each wrote their own Creation accounts: how did anything come to be and then how did they come to be, dealing with the same questions that open the book of Genesis, "Beginnings." Every human culture has asked this question and found its answer. It is a question not of place but of purpose, not the geography of "Where did I come from?" but an existential "Why am I here?" and "Do I have a place in this existence?"

Edward W. H. Vick, in his *Creation: the Christian Doctrine*, writes: "The Christian doctrine of creation is not simply an explanation of the origin of the universe. It holds that God is transcendent and free, that the creatures are contingent and freed, that the ongoing world of history and events in the world are purposive, that within that human history the purpose of Creation is being revealed, that the Redeemer is the Creator. It also teaches that the creation reaches its fulfillment at the end, as the eschaton" (p. 11).

Vick defines cosmology as a theory of the universe and its laws and cosmogony as an account of the generation of the universe.

These two are certainly related, yet their differences must also be noted.

Vick's brief summary of the Christian understanding of creation, above, is similar to mine in Chapter One: God is Creator, Creation is good and a revelation of God, and we have a place and purpose within it.

St. Augustine is arguably the most influential theologian of the classical period, his writing dominating Western Christian thought for a thousand years: from the late 4th Century to the philosophers of the Renaissance and the theologians of the Reformation. In his *The Meaning of Genesis*, he supported the instantaneous beginning of Creation, cosmogony, creation *"ex nihilo,"* out of nothing, or the Big Bang or God saying, "Let there be!" and there was. He also allowed for the development, progress, and even evolution of Creation since the initial event of creation. Augustine spoke of instantaneous cosmogony and evolutionary cosmology.

Augustine viewed the opening chapters of Genesis as containing much allegory and understood the "six days" of Chapter one as describing categories of Creation rather than a calendar of Creation.

Earlier, Origen had described Creation as God bringing order out of chaos. Genesis 1:1-2 says, "In the beginning God created the heavens and the earth. The earth was a formless void and darkness covered the deep and the spirit of the Lord hovered over the waters." This does not describe "creation out of nothing," as there is a formless void and a deep darkness and an earth over which God's spirit hovers. It is a raging chaos, waiting to be tamed and ordered. God gives light, creates life and orders the chaos. Throughout the Old Testament, the Hebrew verb for "create" is applied only to God. Humans "make;" only God "creates."

Both Augustine, the father of Western Christian thought, and Thomas Aquinas, his late Medieval counterpart in terms of influence, argue that we cannot properly understand the Creator without first understanding Creation, that the study of Creation is an appropriate path to understanding the Creator. The apostle Paul makes this argument in Romans 1:18-21, saying in part, "God's power and nature have been clearly seen and understood through what has been made." Aquinas further reasons that the truth of faith becomes a matter of ridicule if we present as dogma what scientific scrutiny shows to be false. If we do not understand Creation, we cannot understand the Creator; studying Creation is therefore an appropriate avenue to better understand God. Any search for truth, from any beginning point, that is fairly and fully pursued will lead us inevitably and inexorably to God, for God is ultimate Truth.

We must remember that the Bible is not a scientific textbook and that it was written in a pre-scientific culture and therefore it is not legitimate or logical to read back into it our scientific mindset. To do so is not fair to either the Bible or the scientist. Vick quotes Paul Tillich as saying, "Science can conflict only with science, and faith only with faith; science which remains science cannot conflict with faith which remains faith .... The famous struggle between the theory of evolution and the theology of some Christian groups was not a struggle between science and faith, but between a science whose faith deprived (us) of (our) humanity and a faith whose expression was distorted by Biblical literalism" (Vick, p. 97).

How did the people of pre-biblical times encounter and understand the divine presence? They did not have the pages of the sacred Scripture as reference, for they had not yet been written. They could not read about God; they had to experience God. "O Lord, our Lord, how majestic is your name in all the earth! Your glory is above the heavens!" (Ps. 8:1) can only be sung by someone who has experienced God in the earth and witnessed God in the heavens. "When I consider your heavens, the work of your fingers," is the utterance of one who has seen sun and moon and stars and seen there the face of God revealed, as Creator seen in and through Creation.

All primitive peoples have described the sacred as surrounding them, the holy as holding their existence, the divine as permeating their daily lives. For their part, the ancient Celts viewed Creation as the first canon and scripture as the second: the scriptures were penned by human hands, however inspired they were; Creation is writ by the very hand of God. What begins as pantheism, seeing a god in every rock and tree, every mound and pond, becomes panentheism, seeing the one true God revealed in and through the works of Creation, the absolute immanence of the transcendent Creator of the cosmos. There is no where that we can go that God is not already there. "If I go up to the heavens, you are there; if I make my bed in the depths of hell, you are there" (Ps. 139:8).

Creation, canon and Christ all reveal the Creator. Science and engineering are no less faithful pursuits than pastor or priest. They are not threat but rather can lead us to praise. As his discoveries were being made public, Kepler as scientist wrote "God is being celebrated in astronomy!" Such knowledge does not lessen faith; it increases praise.

As a mystic, I will confess that God may be known by direct personal encounter, without any intermediary. As a student, I will argue that God may be known by any pursuit of knowledge, from any beginning point, that is fairly and fully pursued, for God is ultimate truth. God may be sought and studied through the natural sciences, for God is nature's God and creation's Creator. As Aquinas said, we cannot understand the God behind Creation and revealed by Creation without understanding Creation itself. We may certainly come to an understanding of God through the pages of the Scripture. As Christians, we declare that Christ is the fullest and clearest revelation of God.

The Christian understanding of Creation or teaching on Creation is that God is Creator and Creation is God's. God is the Tetragrammaton revealed to Moses in the burning bush (Ex. 3), "I am who I am," "I will be what I will be," "I am that which causes to be." Creation is good and God's. Creation is a revelation of God, an incarnation of the divine presence, a revealing of the divine plan. We have place and purpose in God's Creation, to till and tend and to care for God's created order. We are stewards of the earth and responsible for our actions. If, as Vick says, "creation reaches its fulfillment," then there is growth, progress, development, evolution, as part and parcel of God's created order. As Paul writes, "We know that in all things God works for good for those who love God, who are called according to God's purpose" (Rom. 8:28).

A Contemporary Creation Story

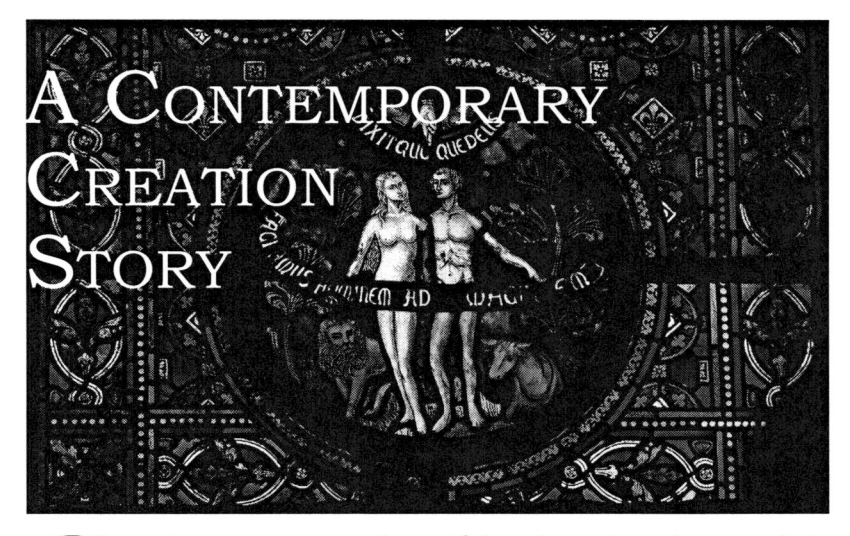

Creation accounts are beautiful and poetic, whatever their source. Genesis One describes a God that is powerful and majestic, ruler over all. Genesis Two describes a God who lovingly dotes over the details of Creation, getting God's hands in the dirt. Psalm Eight stands in wonder and awe in the sheer delight at being alive. Creation stories from other faiths carry the same sense of wonder and majesty, whether stacks of elephants one on top of another, a giant turtle on a limitless sea or star creatures settling on earth. Both faith-based and scientifically based cosmologies attempt to answer three questions: "Why is there something rather than nothing?" "Why is there life rather than just inanimate objects?" and "Why is there consciousness? How does consciousness develop from that which is not conscious?" Faith based creation accounts express a sense wonder that dispassionate scientific analyses cannot hope for, yet the story of Creation from science can be just as wondrous. There is a way to describe the scientific understanding of the creation and evolution of the universe that is both accurate with the evidence and expressive of the wonder. Some call it "The Great Radiance."

Some fourteen billion years ago the Great Radiance occurs, the Big Bang happens. Before, there was nothing; after, there was everything. In one instant Creation occurred. God speaks, and there is the beginning of the universe. There is blinding light and scorching heat with neither sun nor stars. Radio telescopes today pick up the micro-radiation background that is the echo of this Big Bang. What a great, great radiance!

One small fraction of a second later, anti-particles annihilate most particles, creating a cosmos of matter and photons. In theory, the particles and anti-particles, being equal and opposing energies, should have cancelled each other out. For some reason we cannot explain, there were slightly more particles than anti-particles: there was something rather than nothing. The four fundamental forces came into being: gravity, electro-magnetism, and the strong and weak nuclear forces. With these forces came the laws of physics that continue to guide our universe's existence and evolution.

As the universe expanded and cooled over the next 300,000 years, a mere blink in the span of Creation, subatomic particles gradually combine to form the particles of atoms: electrons combined with nuclei to create hydrogen and helium, the first atoms, all floating in the expanse of space.

Then 11 billion years ago the universe differentiated into vast clumps of gas. Amazingly, the universe did not disperse evenly. Minute differences in temperature and mass lead to minor differentiations that were accumulated and accelerated by the law of gravity in random ways. It was a free universe, forming itself freely. These gaseous lumps formed into spheres, in the centers of which nuclear fission occurs. The first stars were born!

A mere billion years later, supernovae exploded as the first generation of stars died, giving their lives in vast spectacular displays of light and radiation. The deaths of the first stars created the complex elements of carbon, nitrogen, oxygen, calcium, iron: the stuff of the world as we know it. Within each of us we carry the remains of these first starbursts; we are made of the stuff of the stars!

Five billion years ago, a massive shock wave occurred of truly cosmic proportion, creating a surge of energy and enriching a gas cloud that would become our solar system and gave birth to our sun, a second or third generation star. The planets began forming themselves along a cosmic disk, outlining their solar orbits with the large outer planets offering shelter to the smaller inner planets as each one finds its place in relation to the others.

Four and one-half billion years ago the infant earth was struck by a cataclysmic collision, carving out the moon from what is now the Pacific Ocean. As the universe gave birth to the solar system and the sun gave birth to the planets, so the Earth gave birth to the moon.

On this active but inanimate planet, 3.8 billion years ago life first occurs. As matter came into being with the Big Bang and there was something rather than nothing, so inanimate matter somehow gave birth to life itself. Molecular chemical reactions have become molecular biological responses. That which is inanimate creates that which is animate! Before there was only existence; now there was life.

Photosynthesis developed about 3.2 billion years ago to cope with an ever increasing shortage of hydrogen and an excess of carbon dioxide. Plants evolved in a way that converted carbon dioxide to oxygen. The earliest life faced a crisis that threatened its existence and creatively found a solution! 2.8 billion years ago the excess oxygen released by photosynthesis created the Oxygen Crisis until life found a way to breathe this noxious, poisonous, volatile gas. Life again found a way and developed respiration in individuals that used the oxygen and created an ozone layer in the atmosphere to protect the planet and its developing life forms. A hydrogen shortage created a solution that created an oxygen abundance that threatened all life. Animals breathe in oxygen and exhale carbon dioxide; plants take in carbon dioxide and release oxygen. The two work together, in balance and harmony. That solution led to a nitrogen shortage. These primitive life forms found a way to fix nitrogen to hydrogen, binding them together, ending the shortage and as a by-product allowing for the possibility of larger, more advanced forms of bio-mass to grow.

Sexual differentiation began about 1.5 billion years ago, spawning a greater diversity among living things. This diversity allowed for greater evolutionary experimentation in a universe free to explore and discover ways of being. Diversity also allowed greater protection from annihilation. Even if entire species became extinct, life in some form would prevail.

Although the creation of the universe began almost 14 billion years ago, the creation of life on this planet did not occur until 500 million years ago. On the cosmic scale, we are almost in the present era before we have life forms large enough to leave evidence of their being! There was a vast explosion of life with massive cycles of experimentation, expansion and extinction taking place as life gave birth to life.

About 245 million years ago a catastrophic event terminated ninety-five percent of all the species on the planet. 65 million years ago another massive extinction occurred as a massive meteorite crashes near modern day Mexico, carved out the Yucatan Peninsula and created an explosion a thousand times greater than all of our nuclear weapons combined and ending life for every creature over fifty pounds. Tiny primitive mammals rose to the occasion and begin an amazing race of expansion and development to fill this void. Again, life found a way!

1.4 million years ago primitive humans began their work of caring for creation with their first domestication: they discovered how to manage and control the gift of fire, of heat and light and a mastering of the world around them. Perhaps 100,000 years ago we developed symbolic communication, making abstract thought possible, allowing us to pass knowledge from generation to generation and creating primitive language and our first sense of culture. They were finding place and purpose in creation!

About 70,000 years ago our ancestors began the Great Walk, leaving Mother Africa to eventually migrate around the globe. They walked their way around the planet, entering the Americas perhaps 15,000 years ago. 10,000 years ago they learned how to farm and domesticate animals. They could regulate their harvests, flocks and diets, their travels and migration, their lives and societies. Civilization became possible. 5,000 years ago we invented the wheel. 4,000 years ago we wrote the first alphabet. We could explore, identify and share our inner world of thoughts and feelings as well as our outer world.

The so-called Axial Age began 3,000 years ago: Homer and the beginning of Greek philosophy, Zoroaster and the Persians, Moses and the Hebrew prophets, Confucius and Lao Tzu in China, Buddha in India. Religion literally "bound us together," helping us make sense of the world we lived in, what it meant to be human and what our purpose and place was in creation. Judaism understood history as a revelation and a progressive journey, from somewhere and toward somewhere, from slavery to freedom and from chaos to order. There was a unity in this universe and the beginning of monotheism. 2,000 years ago Jesus told us we are all children of a loving God and invited us into a new and living 'Way.'

There is something rather than nothing. There is life from inanimate existence. There is consciousness and intelligence from mere existence. Consciousness, God as "that which causes to be," calls forth Creation. That Creation grows, develops, and evolves so that it may respond with a created consciousness, so that we, that through our senses existence may experience existence and through our voices Creation may give glory to the Creator. In the words of Crosby, Stills and Nash and the Woodstock generation, "We are stardust, we are golden, and we've got to get ourselves back to the garden!"

The universe is uniquely formed in such a way that it nurtures and supports our existence. We are made in such a way that we may behold and marvel at the wonder of our existence. The entire development of the universe and progression of Creation has led to this particular moment. Amazing!

God, the great I am, speaks Creation into existence. Creation lives and grows and develops. We emerge, made of the stuff of the stars, with the primordial waters flowing within us, as the consciousness of the cosmos giving voice to all creation, "O God, our God, how majestic is your name in all the earth!"

Evolutionary Biology and the Eighth Day

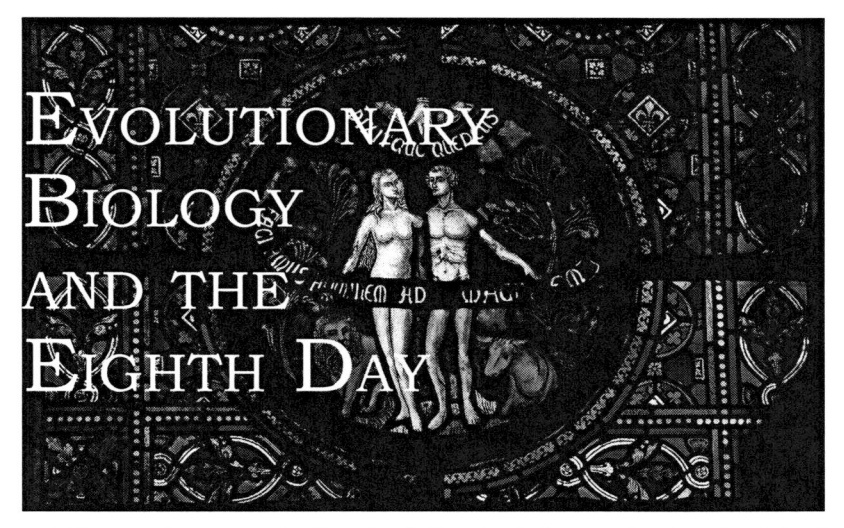

On the seventh day God finished the work that God had done, and God rested on the seventh day from all the work that God had done" (Gen. 2:2). God rested from the work that God had done thus far. The Creation account with which we are most familiar, the account that begins the book of Genesis and the Bible, that recounts six days of creation, says that God rested on the seventh day. Again, it does not say anywhere that God quit!

My father, also a Reverend with over forty years' pastoral experience, has always been comfortable with evolution as a tool in the process of God's Creation. He has a very short synopsis of the creative process: first there was uncreated immaterial consciousness – God; then there was unconscious created material in the universe; and finally, there was created consciousness in humanity. This describes the process thus far.

I remember a Far Side cartoon that pictured a spiral stairs around a mountain peak. On each step was, in succession, a tadpole, a salamander, a big-eyed monkey, a Neanderthal, and a modern homo sapien. From above came a voice, "What makes you think you're the last?" If evolution says anything, it is that Creation is a process, with no reason to believe that the process is over. Biblically speaking, Creation is an ongoing, evolving process from the Garden of Eden to the Garden of Paradise. We are on the way, but we are not there yet!

Paul quotes from Greek philosophers in his message at the Areopagus (Acts 17:22-34), in summary, "God made the cosmos and all throughout it, visible and invisible, and gives life and breath to all within it. God made all nations and allotted boundaries in time and space, so that all might seek God and find God, for God is never far from any of us. Indeed, in God we live and breathe and find our being." This unites Creator with Creation, offering up Creation as evidence of Creator, including not just the product of Creation but also the process of Creation. Evolution is part of that process. It shows a God that is not aloof and distant from Creation, but one that is intimately involved in it and connected to it, so that Creator and Creation are inseparable from one another: "in God we live and breathe and find our being."

In his acclaimed *Evolution's Arrow,* John Stewart, an evolutionary theorist, argues that evolution is directional and progressive and evolution itself is an evolving process. Evolution organized molecular processes into cells, cells into organisms, organisms into plants and animals and animals into herds and societies. Amino acids became RNA which became DNA. Evolution organized chemical processes to become biological processes, becoming in turn psychophysiological and then finally sociological and global. The process of evolution is again evolving, this time through us, organizing local societies into global societies and humanity into a more global consciousness, so that we may in turn be prepared for the next step in the evolution of our species. Evolution's arrow has guided a process from chemical to biological, from molecular to cellular, from organism to organization, from micro to macro, from sub-conscious to self-conscious, to psycho-social, spiritual and now global.

Beatrice Bruteau also writes about the evolution of evolution in *God's Ecstasy: The Creation of a Self-Creating World.* Bruteau is a Chardin scholar with degrees in Mathematics and Philosophy. Drawing from a wide array of examples from the natural and social sciences, she describes the universe as an expression of divine ecstasy as well as divine revelation and describes a universe that is expanding in beauty and increasing in complexity. The second law of thermodynamics argues that, given time, differences in a closed system will even out, entropy will result in the loss of available energy. The universe is generally understood to be a closed system, yet in fourteen billion years we are still expanding and creating. However, the universe has not been, and perhaps cannot be, proven to be a closed system and in some ways appears to be an open one. If the universe is self-creating, then it must be an open system. In an open system, the future cannot be pre-determined because there are too many unknown variables. A self-creating universe is one where we are free to explore and are invited to participate in the development of the universe and intentionally share in the evolution of our species and co-creators with the divine in a process she calls "Christogenesis."

Rupert Sheldrake has a Ph.D. in biochemistry and has taught at Harvard and Cambridge. In his writings, he makes a case for his morphic resonance theory of evolutionary biology. As Carl Jung supported the concept of a universal subconscious memory among humans, Sheldrake supports a collective consciousness among all life forms, giving evolution a purpose and a direction, as well as a will and intention in its development. He replaces the mechanistic view of creation with morphic resonance: past forms and behaviors of organisms influence the present without regard for space or time. As an example, rats in laboratories in New York learn the same maze faster than those in London, and those in New Mexico learn that maze still faster. Each successive group seems to have gained from the previous group's experience.

Sheldrake's theory of morphic resonance has also been compared to physicist David Bohm's theory of Implicate Order. The idea is that, as individual organisms have individual memory, so organizations of those individuals also have a collective memory on an organizational level.

Dean Hamer, geneticist and author of *The God Gene: How Faith is Hardwired into our Genes,* and Francis Collins, a physician and geneticist and author of *The Language of God,* both worked on the Human Genome Project. Their findings convince them that we are genetically designed to be believing beings, to be creatures of faith. Our DNA convinces us that there is 'Something More, Something Else,' something out there or deep within that is divine, sacred, and holy: a God or the gods or the divine spark. This does not prove that such a God exists; it does prove that we want, hope and believe it is so. We are creatures of faith. Neither Hamer nor Collins claim to have found biological or genetic evidence for God, only belief in God.

In *Mere Christianity*, C.S. Lewis does contend that our desiring something, our wanting to believe in something is an indication of that something actually existing. It is not proof, but it is a positive indication from biology of something beyond biology. Indeed, the object of our desire does already exist in our imagination.

More recently, we have psychologist Richard Beck's *The Authenticity of Faith: the Varieties and Illusions of Religious Experience* drawing from William James's *Varieties of Religious Experience* and applying a scientific search through art, doctrine and life experiences in pursuit of authentic faith and using scientific inquiry to move beyond philosophical and theological positions.

We have Hamer and Collins making the case for a created, conscious being wanting there to be an uncreated consciousness. We have Rupert Sheldrake's Morphic Resonance theory of biological evolution describing a collective consciousness, drawing from Carl Jung's Collective Unconscious and David Bohm's Implicate Order. We have John Stewart's theory of an evolution of evolution and Beatrice Bruteau inviting us to join in the divine ecstasy of the Christogenesis of the cosmos. We have Paul arguing that God has planted signs of the divine in all cultures and nations and it is in God that we live and move and have our being. And we have my Dad describing uncreated consciousness, creating an unconscious creation that in turn creates a created consciousness. Sounds like the evolution of creation has just begun!

Toward a Theology of Evolution[1]

We have recently marked the bicentennial of Charles Darwin's birth on February 12, 1809. Perhaps no human has had greater influence on the dialogue between religion and science. His *Origin of Species* and *Descent of Man* were for biologists seminal works and for many religious, books bordering on blasphemy. I fall somewhere between the extremes.

My first encounter with Mr. Darwin was in high school biology class (shared more fully in the Foreword). We were assigned topics for our term papers. Mr. Bishop, my science teacher, suggested I write on the theory of evolution. Mr. Bishop was something of a trouble maker. When he had been a student at my school, he destroyed the chemistry lab with an explosion.

Mr. Bishop knew that my father was pastor at the local Methodist Church. He did not know that dad had been a biology major in college, had copies of Darwin on the shelf and was not going to be anybody's clergy stereotype.

That evening, I broached the subject with my father. I asked Mr. Bishop's fateful question, "What does the Bible say about evolution and creation?" Dad's response caught me by complete surprise, "Which place?" He went on to give me a quick tour of no

---

1 This chapter is adapted from an essay first published in *Sharing the Practice,* the journal of the Academy of Parish Clergy.

less than seven separate and distinct creation accounts in the Bible: Genesis 1:1-2:4; Genesis 2:5-25; Job 38:1-42:6; Psalm 8 and 104, Proverbs 8:22-31; John 1:1-18. He pointed out that each account had its own perspective and purpose and that they were equally true without being mutually exclusive.

Then he got Darwin down, and encouraged me to read that. At fifteen, I was not up to the task, but I learned then that science and religion are not necessarily opposed to one another and that a critical, necessary element for each of them is to keep an open mind, to be a willing student. Biology can be an appropriate entry to theology, and science and religion can accompany and support one another. I also learned that, if we understand God to be Ultimate Truth and Ultimate Reality, then any pursuit of truth that is sincere and thorough must lead inexorably to God. Theologians, philosophers and scientists have all suggested that studying Creation is an effective way to learn of the Creator, with evolution as one of God's tools.

Humility and openness are necessary attitudes in both science and religion. Neither discipline has all the answers in their chosen field. Science cannot explain everything in the universe and we poor theologians cannot claim to know all of God. Too often arguments between the two disciplines result from "either–or" positions that have been taken. My experience is that God is often in the paradox, the "both-and" rather than "either-or". A "both-and" God is a grander, greater God than would be an "either-or" God.

We learn as we grow. Hopefully, my understanding of divine reality will be fuller when I am 60 than it was when I was 15 As I mature, my knowledge expands and matures. As we develop as individuals and as a civilization, should not our understanding of God mature as our experiences of God expand? I am not suggesting that God changes, but that our understanding of God evolves: as our experience broadens, our awareness matures. As spiritual explorers, we should always be open to learning something new about God's presence in our lives.

While on sabbatical, I borrowed a copy of Richard Dawkin's *The God Delusion.* I was looking forward to some challenging arguments, some fresh insight from a different perspective. Dawkins is an eminent scientist, scholar, biologist, lecturer, author and atheist. I was looking forward to some mental aerobics.

I was disappointed.

Dawkins claims that as a well-educated scientist who is an expert in his field, he has no need to study theology or philosophy in order to write about those fields. He makes an error in logic that is far too common: believing that his area of expertise is larger than it actually is.

Scientific knowledge has been described like a chasm: it is certainly very deep, yet it is also quite narrow. Tremendous depth of knowledge does not necessarily lead to breadth of knowledge. Scientific knowledge has proven to be valuable and beneficial. Scientism, which elevates scientific disciplines into religious doctrine, detracts from both religion and science.

Dawkins sets up a straw-man position, establishing his own version of what he is against. Rather than identifying what leading theologians today are saying of their understanding of God, he constructs what he believes those who do believe in God ought to believe. Rather like Rush Limbaugh defining what a liberal is.

I have had the good fortune to teach some college courses, including some on Critical Thinking. Critical thinking is the careful, deliberate determination of whether we should accept, reject or suspend judgment about a claim, and of the degree of confidence with which we accept or reject it. Critical Thinking requires logic, reason and objectivity in order to make an analytical evaluation. It represents the kind of thinking one expects in an open debate or scientific investigation, and it was lacking in Dawkins' book. I'd have given him a "C": it was entertaining, but not constructive.

I wondered where Dawkins got his notion of the divine, but then realized the truth of what he was claiming. He did not get it from anywhere; it was of his own design. He had not studied religion, philosophy or theology; he was just stating his opinion. Reading his book became an exercise in watching a 21st century scientist take on a 12th century philosophy: not on equal terms or speaking the same language or sharing a common worldview.

There are people who do hold to a remnant of medieval theology, but Dawkins does not reference them.

I enjoyed reading the book, but it became light recreational reading instead of a serious intellectual pursuit. I have to say that I do not believe in the God that Dawkins has constructed (only to say that he does not believe in it either). Dawkins says that because he does not believe in the God he has described he is therefore an atheist. I don't believe in the God he describes either, but I am hardly an atheist! I am living proof of the fallacy of his argument.

In brief: theism is the belief in God, deism believes in a God that is Creator but not actively involved in the process of the universe, polytheism is the belief in many gods, pantheism believes all creation is divine and the universe and God are one, panentheism believes the divine permeates creation but there is more to God than the universe can contain, agnosticism suspends belief and admits uncertainty as to the existence of any God or gods, atheism understands existence without any God concept and antitheism is positive disbelief and certainty that there is no God or gods. In this categorization, Richard Dawkins is an antitheist; I consider myself to be a panentheist.

Richard Dawkins is adamant about his atheism and determined to hold to his definition of the God he does not believe in and the faith he does not possess. I remember what my father taught me about the necessity of an open mind in any scientific inquiry. Dawkins seems to be just as extreme a fundamentalist in his atheism as any believer might be in their faith position.

Evidently, the chasm isn't the only thing that is narrow.

If one is open to the theory of evolution as describing a natural process, does that require us to then disavow any divine presence? Must science oppose religion, must biology oppose Creation?

Charles Darwin had studied for the ministry and was preparing for life as a country parson when he took his cruise on the Beagle to study various life forms around the world. He claimed that as he wrote his *Origin of Species* he had as much faith as any bishop. Even late in life he would only say he was an agnostic, not an atheist. Darwin had his issues with institutional religion, as do many others. He enjoyed long walks in the country while his family attended Sunday services. Yet he also saw God as the ultimate law-giver and recognized the moral authority of the Scriptures

An atheist is one who believes God does not exist. An agnostic is one who does not know if or whether God exists. For a scientist, agnosticism can be an expected philosophical position. Science is the critical study of the physical universe by objective observation and controlled experimentation. As such, it is a discipline that is unable to deal with the question of God.

Whereas Dawkins vigorously asserts his atheism, and he is not interested in hearing of any one else's theistic views, Darwin never went that far. He would only confess that he did not have a final answer, and he always regarded the views of religion, of his wife and family, with sympathy and respect. I imagine there are truths which we each do not know. While I am confident in my own faith journey, my experiences and my understandings, I would never make them absolutes for another. I am not god; only God is God.

If God is, according to Paul, the "All in All"(Acts 17:24ff, Col. 1:15ff, etc.), and according to Moses, "That Which Causes To Be"(Ex. 3:14), and the "Without Which Not" of medieval philosophers, then there is no place outside of God where we may objectively observe, no laboratory where we may subject God to our controlled experiments. The discipline of science is simply inadequate for the study of God, and therefore any reputable scientist might admit to agnosticism from the perspective of his or her discipline, just as any theologian would not be able to run a chemistry lab experiment from that discipline.

In fact, any human discipline is inadequate for the study of God. Any God that fits into my brain, that I can claim to know, understand, comprehend, identify, or define is not God. This image would instead be what J.B. Phillips described in *Your God is Too Small*.

I was not raised with the notion of God as a "big guy in the sky", as a tribal chieftain, cosmic storm deity or spiritual vending machine. I was raised on the "Mysterium Tremendum" of Rudolph

Otto's *The Idea of the Holy* and the contemplative classic *The Cloud of Unknowing*.

The God of Dawkins is simply too small, and certainly smaller than the God of Darwin.

In scientific terminology, laws describe what has been observed in certain circumstances and theories propose an explanation of what happens, based upon tests confirming the theory. We call Darwin's idea the "theory" of evolution, and there is overwhelming anatomical and archeological evidence to support it, and we call Isaac Newton's idea the "law" of gravity, even though Einstein's theory of gravity warping the fabric of time-space offers a different perspective. All this is a reminder to be open-minded.

Michael Dowd describes himself as an Evolutionary Evangelist, proclaiming the "Great News" of a sacred world view of cosmic, biological and human evolution. He is an ordained United Church of Christ pastor who travels extensively leading programs that explore and celebrate a 21st century understanding of nature and of God that is large enough to welcome both science and religion, both biology and Creation, both Darwin and Dawkins.

Dowd's book *Thank God for Evolution* includes both the current scientific understandings of the Big Bang Theory and the Theory of Evolution in his comprehension of God and God's revelation in and through Creation. Dowd even includes an essay by Richard Dawkins as an appendix, "Good and Bad Reasons for Believing." Dawkins is more cogent in this essay than in his book. Of course, there are good and bad reasons for either believing or not believing.

Like Dawkins, Dowd presents his understanding of God. Unlike Dawkins, Dowd uses terms consistent with a 21st century, post-modern worldview. In his book, Dowd states that God is "the One and Only Creative Reality that is not a subset of some larger, more comprehensive creative reality. God is that which sources and infuses everything, yet is also co-emergent with and indistinguishable from anything ... the One and Only Creative Reality that transcends and includes all other creative realities" (p. 108). "God is understood foundationally as a holy, proper name for 'the

Wholeness of Reality, measurable and immeasurable'" (p. 110). He offers a list of descriptions: "Source of everything, End of everything, Knows all things, Reveals all things, Present everywhere, Transcends and includes all things, Expresses all forms of power, Holds everything together, Suffers all things, Transforms all things" (p. 110).

Dowd offers enough attributes so that we understand they are descriptions, not definitions, they are inclusive but not exhaustive. Here is a God that is larger than my mind, my intellect, my experience, my existence. A God I can believe in.

Yet Dowd also states, "When I say 'God,' I am not talking about something or someone that can be believed in or not believed in. I'm talking about the Ultimate Wholeness of Reality, seen and unseen – the whole shebang – which is infinitely more than anything we can know, think or imagine .... We don't believe in things that are undeniably real. We know them. We don't believe in water; we are 60-70 percent water. We don't believe in the Universe; we live and move and have our being within this undeniable material and nonmaterial Reality ... This understanding does not reduce the Creator to Creation; rather, it elevates and realizes our sense of divine immanence and omnipresence" (p. 113-114).

Here is a God I can use critical thinking with, a God who can be God of both science and religion. Here is a God that I can take with me when reading *Scientific American* or *Discover* magazine or studying quantum science and chaos theory. Here is a God I can experience in my devotional life, that I have felt in my mystical encounters with the wholly and Holy Other of Rudolph Otto.

Here is 21st century science and 21st century religion, 21st century biology and physics and 21st century creation and faith, conversing and dialoguing together with mutual trust and respect, honesty and humility. Open enough to say, "I don't know it all, I still have much to learn." There is no claim of finality and no attempt to quiet dissension. Instead, there is a request for feedback.

Dowd ends a list of promises given at the beginning of his book by saying, "In the course of reading this book, if anything opens up for you – if you were helped or inspired in some way, or if you found something particularly meaningful – I would love to hear from you. Any and all suggestions for improvement are also welcomed. If you feel one of my promises was not kept, or was overstated, please tell me that, too" (p.xxiii). And he lists his email address.

Along with Dowd's *Thank God for Evolution,* I recommend reading Judy Cannato's *Radical Amazement: Contemplative lessons from Black Holes, Supernovas, and other Wonders of the Universe* and her *Field of Compassion: how the New Cosmology is Transforming Spiritual Life*. Both are beautifully and simply written and serve as opportunities for contemplation and conversation.

It has been many years since I read Darwin's books, yet I remember the impact they had on my adolescent thinking. I have read Dawkins' book and found it enjoyable but not enlightening. I have read Dowd's book and found it stimulating, rewarding, and encouraging: an open and positive argument rather than a closed and negative one. I strongly recommend it to everyone, for he is solid in his science and current in his theology. He ends his text with this question for us to consider, "What would it mean if I knew and felt in my bones that everything is a miracle" (p. 344)?

Creation and the Divine Feminine

We sometimes get trapped by images of our own making. The Bible offers over six hundred descriptions, allusions, metaphors or images for the divine. They are overwhelmingly not human: the mighty rock or fortress, raging sea or storming sky. Those descriptions that are human tend to be masculine, but are not exclusively so. God is the mother bear protecting her young, the mother eagle teaching her young to fly. Jesus describes himself as a mother hen, gathering her chicks around her. These are not weak descriptions. No one wants to get between a mother bear and her cubs. It takes courage and hope for a mother eagle to cast her young upon the wind and dare them to fly. The mother hen gathers her young that she may protect them, even at the cost of her life.

In reading *The DaVinci Code*, I was surprised to see the claim that the Tetragrammaton, the word representing the very name of God and so holy that it dare not be uttered, sometimes written *yhwh*, was actually a joining together of the masculine and the feminine forms of the 'to be' verb in Hebrew. Knowing that this was a work of fiction and that Dan Brown was inclined to alter sources for the sake of his story, I did my own research. I looked *yhwh* up in my biblical dictionary and was surprised to learn that some scholars do believe that this is the correct etymology of the word. Like anything else in Biblical interpretation, there is not universal agreement.

The holiest name of God was a combining of masculine and feminine. This name was only uttered on the Day of Atonement, and only by the High Priest, and only within the Holy of Holies, the innermost court in the temple in Jerusalem. A rope would be tied around the high priest's left ankle so that, if he should be struck down by the divine presence, his body could be dragged out of the Holy of Holies. This name above all names may well be a combining of both masculine and feminine.

I have always thought of creation as more of a feminine act than a masculine act. As a man, I can make objects. I can saw and hammer, cut and assemble. I can make, but I cannot create. I can not take from my substance to create substance, from my person to create another person. Every mother does this with every birth: from her person, a new person; from her substance a new substance; from her essence a new essence.

I know that God is more than male or female, more than any amount of descriptions I can write and other than all the descriptions I can write. God is the holy other and the wholly other. Yet I try to describe what I have experienced. And I have experienced the act of creation as more of a feminine act than as a masculine act.

Creation is the greatest miracle that God can perform. I may be taken to task, but I would argue that creating the universe is a greater act than saving it is. Creation is the unique work of God, the One who causes to be, and yet every mother experiences it, not by hearing about it or reading about it, not studying it but being it. Every mother creates, causes to be, gives life.

Nurture is likewise traditionally seen as feminine and nurture is an innate part of nature. Nurture is necessary to bring created beings into their own full independence. We've all seen animal mothers provide nurture for animal infants who are not their own, even not their own species: cats raising bunnies, dogs raising lambs, etc. The need to receive and provide nurture seems to be part of our being.

Meister Eckhart raised the question, "What does God do all day?" and offered this answer, "God lies on a maternity bed, giving birth to creation all day." The act of creation may be the fullest expression of the feminine spirit.

Quantum Physics and the Dance of the Cosmos

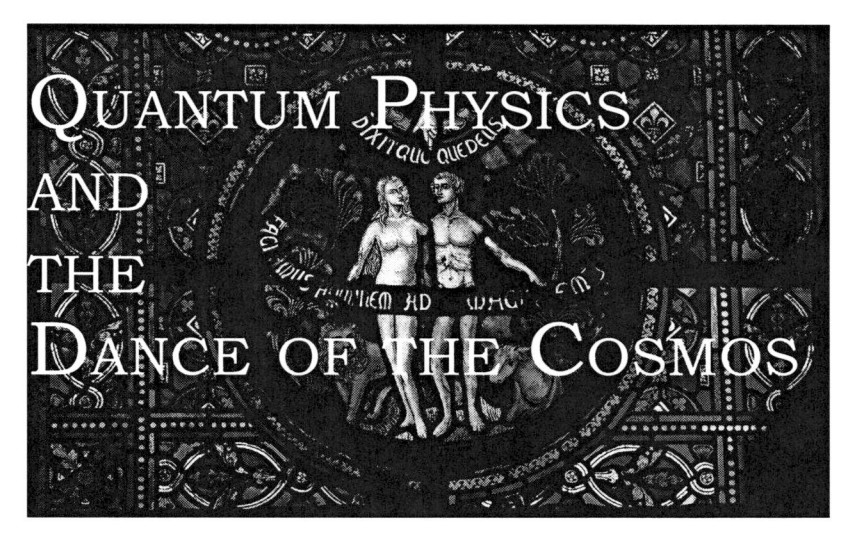

In 1905 the scientific world thought they were almost finished. They had answered all the big questions and most of the small ones. Knowledge of the physical universe seemed to be nearly complete. Then an upstart young patent clerk published an article on special relativity and Albert Einstein turned the entire scientific community on its head.

That is a slight over-statement, but only a slight one. Other experimental and theoretical scientists were also finding holes in what seemed a nearly seamless garment. Max Planck observed discrete states of energy and that energy is only available in multiples of these discrete states, that multiplier being called "Planck's constant" and these states being called quantum states; Erwin Schrödinger popularized its use, but Einstein's theories of special and general relativity, and the theories of the conservation of energy, quanta of energy, wave-particle duality, zero point energy, etc. founded quantum physics and substantially re-wrote scientific thought in the twentieth century. Werner Heisenberg's uncertainty principle demonstrates that if we determine the speed of a particle we cannot know it's location and if we determine it's location we cannot determine its speed. Erwin Schrödinger's thought experiment called "Schrödinger's Cat" also demonstrates the limits of certainty, in that we cannot know if the cat is dead or alive until we open the box.

Niels Bohr, a contemporary of Einstein, sometimes collaborator and sometimes antagonist, said, "Anyone who is not shocked by quantum theory has not understood it." Going with Bohr's shock value of quantum theory, the Von Neumann-Wigner interpretation of Schrödinger's thought experiment is that the cat must be both dead and alive as long as the box remains closed. We cannot know until we observe, just as the photons in the double slit experiment can be anywhere until we observe and measure them. Everything exists in potentialities until our observation causes them to specifically locate in reality, consciousness causes the collapse of the wave function of potentiality into the particle function of observed actuality. The idea of a cat being both potentially alive and potentially dead certainly fits with Bohr's description of quantum theory being shocking!

I am neither a quantum physicist nor a scientist. I have read a dozen books on quantum physics but have no illusion about understanding it. Though leading the scientific world in the late twentieth century, it was not taught in my high school physics class. The concepts formed and questions raised by quantum physics are truly mind boggling.

What is life like in the quantum world? How are matter and energy interchangeable? How does energy create matter? How do we understand the immediate interconnectedness of everything? How can we live with confidence in a world without certainty but only probability? What are these stings and super-strings, membranes and dimensions? How many universes exist? How can we know? What holds things together? What keeps things apart? What is dark energy and dark matter? How can we accept that any of these answers are complete?

We begin by looking at the universe in the smallest possible units, beneath objects to molecules, beneath molecules to atoms, then into the sub-atomic realm of quarks, bosons, muons, photon, leptons, etc. where particles might be up or down, top or bottom, strange or charmed. The smallest possible particles may be equally well described as both wave and particle and yet neither wave nor particle, as much a potentiality as a reality. Energy is the basis of the entire universe. Einstein's theory of $E=mc^2$ offers a formula for the conversion of energy into matter and the conservation of energy within the universe. As energy, the particle is in constant movement. The movement of one particle of energy vibrates in relation to the particles around it. These vibrations form a harmony on the sub-atomic level that in large enough numbers forms the appearance of material existence.

We know that every bit of matter: the book within your hand, the chair upon which you sit, the table at your side, is composed of molecules and of atoms. We know the description of an atom is much like that of our solar system: there is a large nucleus at the center that contains most of the mass of the atom. Around the atom, bouncing in and out of probability zones, are highly charged electrons that have very little mass. Most of the space of the atom is just that: empty space. If most of the atom is space and atoms form the structure of our books and tables and chairs, and all else in our material universe, then most of the seemingly solid universe is actually space!

On the smallest subatomic level, the smallest particles we can detect are really not matter at all, but energy particles dancing in harmony with one another. When taken together, these tiniest of all particles form the grandest of all dances, a cosmic dance that creates the universe and all that is within it.

The principle of Zero Point Energy states that energy is a positive effect throughout the universe. For instance, even at absolute zero, when theoretically all energy has been sapped from the system, liquid hydrogen will not freeze under normal atmospheric pressure; it still has residual positive amount of energy, its Zero Point Energy. So it is throughout the entire universe: even in the vast reaches of seemingly empty space, there is Zero Point energy, there continues to be motion at the sub-atomic level, the cosmic dance continues.

Radio telescopes need to be adjusted for the existence of cosmic microwave background radiation, the relic radiation from the Big Bang. Fourteen billion years after this primal cosmic explosion, we still have the echo of God's mighty "Let there be!" radiating throughout the universe.

Quantum entanglement suggests that all parts of the universe, in their farthest reaches, may be connected. It is as if a gymnasium were filled with ping pong balls. If we were to push on one of those balls, there would be an immediate reaction throughout the entire room. Expand that gymnasium to the size of the universe and imagine that each particle within it is interconnected with other particles, so that communication among them is immediate in spite of distance. Each part of the whole exists in relationship with other parts of the whole to create one vast cosmic unity, inter-related and interconnected. Our sense of separateness is an illusion, for on the smallest and the grandest levels, all the universe is united as one.

The Heisenberg Uncertainty Principle puts us in a world of probabilities rather than certainties. We realize that the act of our observing an experiment affects the outcome of that experiment. We cannot be detached observers; we are instead additional subjects participating in the outcome of the experiment we attempt to observe. In particle physics, it means we can measure the speed of an electron or the location of it; we cannot know both. Measuring one impacts the measurement of the other. The subatomic world of particle physics, where quantum physics reigns, is a world where nothing is certain. Complementary variables cannot be simultaneously defined. We must live with confidence in a world without certainty, perhaps a new expression for what we have traditionally called "faith."

The electrons whirl in a probability cloud, where the densest parts of the cloud represent the highest probability for the electron, but the electron could be anywhere; we can never be certain.

Experiments with photons, the subatomic particles of energy that create light, demonstrate that photons behave both as particles and as waves, depending on the details of how we study them. Both-and, not either-or. Imagine a world where we can choose both-and and not just either-or. Imagine a God that might be both-and rather than either-or. Such a God would be expansive and inclusive, inviting of opposites and welcoming of options. It might be possible for us to disagree and yet both be right, or both be wrong!

Einstein's Theory of Relativity expands to state that everything is relative: energy and matter as we have seen, and time and space and gravity. Newton's apple and the planet earth exist relative to one another. The mass of the apple is considerably less than that of the earth, so we see the apple fall to the earth. The earth is also affected by the apple and, though immeasurably small, it shifts to the mass of the apple. As objects move through space, they bend the fabric of space as a bowling ball impacts a trampoline. Time is related to speed. As I approach the speed of light, time slows down for me, though not for any who might observe me. If I could take a space ship to the speed of light and back, I would not age as much as those left behind, nor would I be as aware of the passage of time; my time would be different.

The expansion of the universe has long been a curiosity for astronomers. Classical physics had predicted a stable universe, one that is neither expanding nor contracting. When it was demonstrated that the universe was indeed expanding, scientists could calculate back to the probable dating of the Big Bang. Is the rate of acceleration sufficient to keep the universe expanding indefinitely? In which case it will grow colder due to a constant amount of energy being dispersed over a greater area. Is the rate of acceleration slowing so that the universe will collapse in a big crunch?

Edwin Hubble, after whom the telescope was named, discovered that the rate of expansion in the universe is actually increasing! This is demonstrated in what is called the red shift. The speed of light is a constant. The light of an object that is approaching us will shift toward the blue end of the spectrum; the light of an object that is receding from us will shift toward the red end of the spectrum. Hubble was surprised to discover that, measuring in any direction, the light of every object is shifting toward the red. This red shift means that everything in the universe is speeding away from everything else at an increasing rate of acceleration.

To explain how shocking this is, imagine firing a gun. The bullet accelerates to its greatest speed when it leaves the barrel. That rate of speed decreases with time and distance until the bullet falls to the ground, having spent all its residual energy. For the bullet's speed to increase, it would need to have its own supply of energy, as it does when in the barrel from the explosion of the propellant.

All the known energy in the universe that we can calculate isnot sufficient to account for this increasing rate of acceleration, yet it is there. How can this be? This extra, unknown energy is called "dark energy." Dark not because it emits no light but dark because, while we believe it exists and it explains what we can observe, we cannot directly observe it. There is a sense in which we must take it all on faith.

Analyzing what we can observe, we measure the stars of the galaxy and calculate their collective mass. The mass that we can measure and calculate is insufficient to provide the amount of gravity necessary to hold the galaxies together, and yet they do not fly apart. How can this be? There must be more mass than we presently know. We call this "dark matter." Matter that must exist because the galaxies stay together, but we cannot prove its existence or directly observe it. We simply believe it to exist.

Analyzing what we can observe, we measure the stars of the galaxy and calculate their collective mass. The mass that we can measure and calculate is insufficient to provide the amount of gravity necessary to hold the galaxies together, and yet they do not fly apart. How can this be? There must be more mass than we presently know. We call this "dark matter." Matter that must exist because the galaxies stay together, but we cannot prove its existence. We must take it on faith.

We estimate that over two thirds of the universe is dark energy and over one fourth of the universe is dark matter, meaning that we estimate that the most we could possibly know of the universe, if we were to know all that is knowable, is five percent. I do not count that as great knowledge, even passable knowledge in any course of learning. I take it as a call for every physicist to be humble, realizing how much they don't know. I also remind my scientific skeptics that it takes a great amount of faith to believe in that other ninety-five percent. Of course, if the physicist can only know five percent of their chosen field that is still infinitely more than any theologian can know of theirs!

Creation in Contemporary Experience

I like to think of dark matter and dark energy as a scientific version of "The Cloud of Unknowing," a fourteenth century mystical classic. The more I enter the cloud, the less I know, until I am fully within and totally ignorant and, with Job, submit in dust and ashes (see Job 42:1-6).

One theory coming out of quantum physics is that we do not live in a single universe but in an infinite number of multiverses, even that each choice made spins off another universe where each choice is taken. All things that can exist do exist in some universe. Rather than demonstrate that God is not necessary, this theory tells me that God is where all things are possible.

Some scientists theorize multiple dimensions. We are aware of three dimensions because we are three dimensional beings. Some might add time as a fourth dimension. There are theories for nine or eleven dimension, where some are so spread out we cannot recognize them and some are so tightly interwoven we cannot experience them. There is an M-field, a dimension that is like a membrane that passes through all things in all dimensions and is what holds everything together. Again, this sounds more like a twenty-first century description of God rather than a rationalization against God. What is it that passes through all things, penetrates all things and yet holds all things together?

Here is Paul describing Christ, "In him all things in heaven ad on earth were created, visible and invisible, in him all things hold together" (Col. 1:15-17). Sure sounds like the M-field!

Stephen Hawking is a brilliant theoretical physicist who has written eloquently in his field. His *The Universe in a Nutshell* and *A Brief History of* Time are deservedly popular classics. With Richard Dawkins, he has argued against the existence of any divinity. In *The Grand Design* he dispatches theology in a few pages, after he has dispatched psychology in a few paragraphs. Hawking's explanation that quantum gravity explains why everything is the way it is begs the question, "Why is there quantum gravity?"

We have another example of scientific knowledge as a vast chasm: incredibly deep, inexplicably narrow. Hawking knows more of physics than I can hope to know, but little of theology. I will confess again how little I know of either. Socrates said the wisest man is the one who knows that he does not know.

Einstein pursued his quest for a Unified Field Theory, or a theory of everything, that would explain all the energy behind everything in the universe. There are four forces at work in the universe: the strong atomic force that holds sub-atomic elements together, the weak atomic force that accounts for radiation, the electromagnetic force that combines electricity and magnetism, and gravity. A Unified Field Theory would explain how these forces worked together. Einstein never achieved this.

He remains arguably the greatest scientific mind of the twentieth century. Einstein said, "Religion without science is lame; science without religion is blind." He said, "Imagination is greater than knowledge," and "It is not that I am more intelligent, only more curious." He said, "We cannot solve the problems before us with the same level of thinking that created them." Einstein regarded himself as an agnostic, but not an atheist. He believed in a pantheistic God, a divine presence in all times and places, but not a personal one. Einstein contended "There are two ways to live in this world: as if nothing is a miracle or as if everything were a miracle. I choose the latter." I have used points from quantum physics to argue for the possibility of God. When people respond by arguing against those points, I just ask them to take it up with Einstein, the greatest scientific mind of the century.

Chaos Theory and the Risk of Freedom

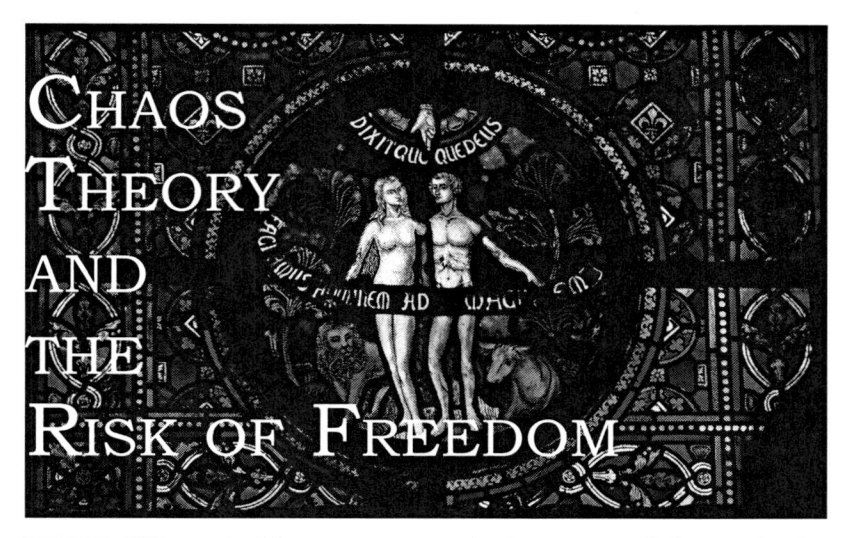

"When God began creating the heavens and the earth, the earth was a formless void, a deep darkness, a raging chaos, and the spirit of God hovered over the face of the deep" (Gen. 1:1-2). So begins the most familiar of all the Biblical creation accounts. It tells the story of God speaking forth the universe by the creative force of God's word. It does not tell of God creating out of nothingness, "ex nihilo." God begins with chaos, and out of this chaos God brings forth all of creation. Biblically, chaos is the seed-bed of creation.

Chaos Theory is a mathematical theory of apparent randomness. It is a theory describing complex natural systems that are so sensitive that small changes can give unexpected results, giving the appearance of randomness that belies a hidden process. These natural systems are fundamentally unstable to even the smallest influence, so that their reactions are effectively random. Heisenberg's uncertainty principle again reminds us that if we determine one parameter we cannot determine another one. It has application to a variety of other fields, including meteorology, physics, history and economics.

Chaos Theory is described in the parable: For want of a nail, the horse was lost. For want of a horse, the rider was lost. For want of the rider, the charge was lost. For want of the charge, the battle was lost. For want of the battle, the war was lost. Chaos Theory is behind the question, "Does a butterfly's wings flapping in China create a storm in California?" Chaos Theory is part of Charles Mackay's classic *The Madness of Crowds*, witnessed in investment bubbles ranging from the tulip mania in The Netherlands in the Seventeenth century to the housing bust in 2008.

Chaos is all around us. We experience it rushing through a traffic light and hitting another car. We experience it when we miss a connection by minutes or make one by minutes. Randomness is all around! We tend to fear the freedom and risk of this randomness; we cower before the chaos, even though it is a good thing.

From chaos, God brings forth creation. From chaos comes order, and order can be detected at work beneath the chaos. From chaos comes our freedom.

Fractals are often used as an example of Chaos Theory at work. Fractals appear to be quite random in their formation, but upon closer examination, minute patterns can be detected and their repetition can be calculated if not predicted. A fractal may be described as an irregular repeated pattern that can be repeatedly subdivided into smaller versions of itself. Fractals are used to describe the irregular patterns of seashores and mountain-scapes. Fractals demonstrate order in the chaos and chaos in the order and how the two work together in creation.

Einstein argued that he could not believe that God plays dice with the universe. I might argue that, in Chaos Theory, God *does* play dice with the universe, but the dice are loaded. Hawking and others have argued for the existence of an infinite number of multiverses and our universe is merely one example. Believing in an infinite number of multiverses seems to be easier than believing in a God who is infinite. Hawking might have unintentionally provided a support for God's omniscience and our free will: God may know all possible outcomes, but not which outcomes we actually choose. I don't know about other universes. I only know that the universe I am a part of is uniquely designed to support human life.

If the Big Bang had been mili-seconds faster the entire universe would have scattered asunder. If the Big Bang had been mili-seconds slower the accumulated gravity would have had it collapse upon itself. If the collision of matter and anti-matter had perfectly annihilated each other, as might have been expected, there would be no matter in existence.

The big three questions of cosmogony are "Why is there something rather than nothing?" "How did inanimate matter give birth to life and animate matter?" and "How did consciousness arise from a creation unconscious of its own existence?" Chaos Theory is a part the answer to each of these questions. Randomness happens. As the chaos mathematician in *Jurassic Park* warned, "Life will find a way."

Freedom required randomness, things and events not to be fixed, and randomness is an expression of chaos. I do not believe in a deterministic universe or a God who has fixed the future or who dictates our destiny. I experience a world where I am free to choose my path. I may make mistakes, I may choose poorly, but they are my mistakes and my choices.

I have experienced God as a God of love. As 1 John records, "God is love." This is part of the essence of who and what God is: God is love. Love exists to give itself in love in the hope that the love will be reciprocal: it will be received in love and returned in love. Love seeks to love. Love requires freedom. If I cannot freely say "No," I can't really say "Yes."

I tell a story that never happened. In grade school I had a crush on Vickie Reed. That much is true. I describe myself chasing her on the playground, tackling her and holding her down saying, "I'm going to make you love me!" The scene is illustrative as well as amusing. Love cannot be forced. No one can make another person love them, and nor can God. That is not the way of love.

An omniscient God may know all the potential choices and all their inherent consequences, without knowing which choices we are going to make, thus preserving both God's omniscience and our freedom. In a closed system, this conscious knowing by God might collapse all the potential wave function into specific particle actualities, but God is the God who lets be. God says to Moses, "I am what I am" or "I will be what I will be" or "I am what will be."

The Hebrew of Exodus 3:14 is difficult to translate with certainty. God is the God who lets be, who causes to be, who keeps the system as an open system.

God is love. As such, God wants nothing more than for us to love in return. For that love to be real, it must be free: love requires freedom. For us to love, we must be free creatures and have free will. Our freedom requires us to live in a universe that is likewise free. So God risks everything in creating a free human on a free world in a free universe in the hope that we may choose to respond to the love of God by which everything has been created. This holy freedom can be described in terms of randomness and chaos.

Freedom is something we give much lip service to. In the movie *Braveheart*, William Wallace cries out with his dying breath "Freedom!" Earlier in the movie, he quotes from the Arbroath Declaration (actually written fifteen years after his death in 1305), justifying his rebellion against Edward I, "It is in truth not for glory nor riches nor honors that we are fighting but for freedom – for that alone, which no honest man gives up but with his life." The Arbroath Declaration is seen by some as a precursor to our Declaration of Independence; it was signed on April 6 1320, a date officially recognized by the U.S. Congress as "Tartan Day."

Freedom is clearly something we aspire to, yet it is also something we flee from, as Erich Fromm described in his classic *Escape from Freedom*, sometimes titled *Fear of Freedom*, written on the verge of World War 2. There is negative freedom, freedom from, and positive freedom, freedom to. Freedom offers us the opportunity for authenticity, to fulfill ourselves, but by definition this comes without rules and regulation. The risk of freedom and the challenge of authenticity can seem to be too much. Fromm sees that we often substitute authoritarianism, destructiveness or conformity for our freedom, meaning that rather than being free creatures in a free creation we choose not to be ourselves. As Benjamin Franklin said, "Those who give up freedom for security deserve neither."

In the movie *The Adjustment Bureau*, we see a dramatization of the struggle between free will and pre-determinism. There is a question about how free we really are, with the reply that we get to make the small choices while the Bureau makes the big questions. An example of a small choice is choosing our own toothpaste; an example of a big choice is choosing our spouse. In the end, they are given a clean slate and an uncharted future, free to determine their own destiny.

I am convinced of the love of God and that love requires freedom. I confess that freedom carries risk.

One December afternoon a beautiful young high school girl was driving to the county hospital to work as a volunteer, a "candy striper" because her uniform was pink and white striped. She was popular, a cheer leader and an honors student; in a word, she had everything going for her and a bright future ahead. It was snowing and blowing and the road was ice covered. She lost control, spun out, ran into the ditch and hit a telephone pole. This beautiful young girl who was on her way to do something good for others was killed at the scene.

At the visitation, I heard people offering comfort to the family. "It was God's will." "Heaven needed another angel." "It was her time." I took a deep breath and choked down my theological rebuttal; it would offer no comfort in this time and place. I do not believe in a God that kills people. That is not how I have experienced God or how I understand God. Her death was a tragedy, but not an expression of God's will.

For God to have taken that young girl's life, it is not enough for God to have set that storm into place. The ancient storm god of Psalm 18 can do that. God would have had to set that highway on its course winding through the countryside so it twisted and turned. God would have had to built the highway so that it had a crown and the car could slide into the ditch. God would have had to kept the highway department from keeping the road surface dry so that it would be icy and slick. God would have determined her route, her rate of speed, her attention to her driving. God would have had to plant that telephone pole in that precise spot, not two feet away in any direction, so that the car could hit it just right.

For God to have taken that young girl's life, we would all have to be puppets on strings, incapable of any free will and therefore not responsible for any of our actions. That is neither what I have experienced nor how I understand God and the universe to work. Martin Luther King, Jr. said, "The arc of the universe is long, but it bends toward justice." I would add that it also bends toward love.

There is chaos in the universe; there is randomness at work among us. This sets us free: free to choose, free to be, free to become, even free to love. God says the love is worth the risk. "For God so loved the world ("cosmos") that God gave God's most beloved Son, that whosoever believes, might be saved."

Caring for Creation: On a Christian Ethic

If we see all creation as God's gift and God's property, even an expression of God's self-revelation, how do we demonstrate that belief through our behavior? How does our experience of God in creation affect our ethic toward creation? What insight from evolutionary biology, quantum physics and chaos theory can be applied to our contemporary understanding of Christian ethics? In a word, what difference does it all make?

Genesis 2:15 states "God put the human (human/humus, earthling/earth, adam/admah) in the Garden to care for it and look after it, to till and tend it. Biblically, we are placed here to look after what remains God's garden. We may be formed by God, even in the image and likeness of God, and live by the breath of God within us, but we are also from the soil of creation as much as the soul of the Creator, formed an earthling of the earth, a human of the humus, the dirt around us. We are formed from the earth and our bodies return to the earth. We are in no way above or detached from creation, but rather very much a part of creation. And we are formed and charged to be caretakers of God's garden, stewards of God's creation. The Bible begins in the Garden of Eden and ends in the garden of Paradise; from beginning to end it is all God's garden, entrusted to our care.

The Hebrew in Genesis 1:28 that is sometimes translated "subdue and dominate" the earth could just as well be translated "control and rule" or "manage, govern and direct." It does not mean "exploit, exhaust, abuse or devastate." It can well be argued biblically that "being green" is God's idea.

We are not here to plunder the earth, to take from it indiscriminately, to use and abuse the earth, to devastate God's garden. We are here to care for it. Being a steward means we provide personal care for what is common property. The word "steward" comes from the Anglo-Saxon "sty-ward," the keeper or guardian of the pig sty, which represented the accumulated wealth and the future welfare for the tribe.

There's a story of an avid gardener who was so proud of the fruit of her efforts. She was giving a tour of her extensive landscaping when her friend commented, "My, how wonderfully God has filled your yard!" The gardener replied, "You should have seen it when the Lord had it to himself."

We are co-creators, participating with God in the ongoing work of the evolution of creation, assisting in the divine work of bringing order to chaos. Progress is possible and positive. Progress is always change, but change is not always progress. As God's co-creators and caretakers, we manage the earth's resources, we till and tend God's garden, but always with the mindset that this is divine work and divine property and divine blessing. It is not ever just about us!

James Watt, devout evangelical Christian who was Secretary of the Interior under President Reagan, believed in the imminent return of Christ. Therefore, he believed, we could exhaust the world's resources without concern for the future, since there would be no future to worry about or plan for. So he thought. Unfortunately for this scenario, every prediction of the return of Christ at the end of time throughout all history has proven wrong. Granted, this is an issue that we need only be right on once, but it clearly is not what we should expect for our "plan A." We are charged, in Genesis 2:15, with the responsibility of caring for and looking after creation, tilling and tending God's garden as God's chosen stewards, with no certain knowledge of the end of our responsibility or the termination of our charge. It is not our place to attempt to force God's hand.

My theory is that we need to care for this planet at least until we can find another one to use up! There is no "away," so we can't throw things there: everything must be someplace, it is a fact of our material reality. We are called to reduce, reuse, recycle and repurpose as best we can.

As the Native Americans say, we do not inherit the earth from our parents; we borrow it from our children. I want my grandchildren to be able to play in the grass, breathe the air and drink the water; this makes me a conservative in the truest sense of the word. Caring for the planet is an essential part of our contemporary Christian ethic.

Bob Edgar, former US congressman and General Secretary of the National Council of Churches, wrote *Middle Church: Reclaiming the Moral Values of the Faithful Majority from the Religious Right* (Simon and Schuster, 2006). In this book, he argues that the three essential principles of a Biblical ethic are caring for the planet, providing justice for the poor and working for peace. Planet, poor and peace.

One of the Bibles on my shelf is *The Poverty and Justice Bible* (American Bible Society, 2008). It highlights in orange every verse in the Bible that has to do with social justice, over 2,000 verses. There are very few places where you can open it up and not find some verses highlighted on the two pages before you.

Richard Stearns joined World Vision as their president after decades of successful business management and what he thought was years of faithful service as a Christian. He wrote *The Hole in our Gospel: What does God expect of us? The answer that changed my life and might just change the world* (Thomas Nelson, 2010). Over half the world's population cannot find adequate work, food or housing. This would at least imply that it is wrong for any child in North America to have an electric toothbrush until every child in Africa has a glass of milk a day. It is not right, ethical, for us to value our wants or desires above another's needs or necessities.

Another study determines there is sufficient food produced to feed the world's population, if we would distribute it more equitably. Archbishop Helder Camara said, "When I feed the poor I am called a saint. When I ask why the poor have no food I am called a communist." In one week the nations of the world spend on arms and armament, that no one hopes to use, enough to feed the hungry of the world for a year.

The hot issues that have been so much a part of Christian political action, abortion and homosexuality, are barely mentioned in the Bible. Abortion is nowhere mentioned. There are a few verses in Leviticus and a few in Paul's letters that mention homosexuality, but Jesus nowhere mentions it. I was once sitting with a fellow pastor whose denomination was greatly divided on how to relate to gay or lesbian Christians. He expressed concern with the verses in Leviticus about it being an abomination. We were sitting enjoying some pizza for lunch. I reminded him that our eating pork was an abomination, our mixing meat and cheese in one dish was an abomination and our wearing blended fabric was an abomination, yet we casually gloss over these other noted abominations without giving it a thought. Our concern with homosexual acts is not based on scripture but ourselves, unless we also avoid these other abominations: eating pork, eating shellfish, mixing meat and milk products together (no more Hamburger Helper) and wearing blended fabric, even disrespecting or disobeying our parents, all equally abominations. I had a United Methodist bishop confess to me that he believed the reason we are uncomfortable with homosexuality is because we are uncomfortable with human sexuality and our own sexuality.

These current hot political topics may well be a matter of ethics, but they are not important in the Bible. What is important Biblically is our care of the planet, our care for the poor and our working for peace. These three principles are of profound importance in the pages of the Bible.

There is another principle that is of more importance to Christians because it is of more importance to Christ: love. Jesus says the first and greatest commandment is to love God with all our heart

Creation in Contemporary Experience 51

and soul and mind and strength, and the second commandment is like unto it, it makes the first commandment tangible: to love our neighbor as we love ourselves. And who is our neighbor? Jesus responds with the Good Samaritan: our neighbor is anyone who is in need. The night before his death, Jesus said to his followers, "Love one another, even as I have loved you, so you love one another. Greater love has no one than this: to lay down his life for his friends" (Jn. 13:34-35, 15:13). This how we prove the truth of the gospel and the integrity of our faith: by loving one another. Jesus even tells us we are to love our enemies! Paul makes love the greatest gift and the first fruit of the Holy Spirit. Loving one another fulfills the law and the prophets. It can well be argued that the gospel in one word is love. Live in an attitude of love with ourselves, with those around us, with the entire world. As St. Augustine said, "Love God and do as you please," because if we truly love God, we will want to do what pleases God as well as ourselves. Love everybody – no exceptions!

Jesus teaches that it is in our love for one another that we prove the integrity of our faith: how be behave shows what we really believe (Jn. 13:35). How we love shows the truth of what we believe. In the early years of the church, Greek and Roman philosophers laughed at the silly notions of these strange Christians, how they believed in a crucified savior. As Paul writes to the Corinthians, a crucified Christ is shameful to the Jews and foolish to the Greeks (1 Cor. 1:22-30). The one witness that Greeks and Romans could not argue against was "How these Christians do love one another!"

Our Christian ethic in one word is love: love of the planet God entrusts to us, love toward all our fellow humans, near and far, known and unknown, even our enemies, who are all our brothers and sisters in Christ regardless of their religious or political beliefs, love to our fellow inhabitants and residents on this planet, both plant and animal, love that is unconditional and unlimited, love that represents and incarnates the love of God, the love that is God.

We are free. We may choose not to be Christian. If we choose to be Christian, we are called to love. For Christians, loving is not a choice or an option, it is a command. If we are Christian, we must love God, love one another, love our neighbor, love ourselves, and love our enemies.

In quoting Deuteronomy 6:5, where we are called to love God with our heart and soul and strength, Jesus adds loving God with all our mind. The intellectual love of God is part of this great commandment. Our faith is to be a reasoned, sensible expression. As my United Church of Christ says, "our faith is 2000 years old, our thinking is not!"

When Jason was in second grade he had an assignment to write about the three necessities of life: food shelter and clothing. Jason put his own unique spin on it. He said the three necessities were not food, shelter and clothing. Clothing was not a necessity. If we wore clothing to stay dry or to keep warm, it was really portable shelter. If it was warm and dry, we could survive fine without any clothes. He said instead of clothing, the third necessity for life was love. Food, shelter and love are the three necessities of life!

Another UCC quote: "To love is to care, to care is to do."

If we take seriously the three principles expressed throughout the pages of scripture, caring for the planet, caring for the poor and working for peace; if we take seriously our Lord's command for us to love, then this must be borne out in our lives. Our behavior must model what we believe. Our walk must match our talk.

The Morphic Field theory of evolutionary biology tells us we are all connected: the world, the universe and all creation are all parts of one great cosmic whole. Like quantum entanglement, the M-theory of quantum physics posits that everything in the universe may be intimately and immediately inter-connected. The globe is one community. As my wife says, we are all members of one race, the human race. Gandhi declared "All men are brothers." I would add, "All women are sisters." Our contemporary experience of God tells us that we are all members of one cosmic community. As *Desiderata* says, we are all children of the universe. The fractals of chaos theory remind us that the smallest conceivable action may have the greatest possible reaction; the smallest cause may have the greatest consequence. Everything changes, everything is in a state of flux,

Creation in Contemporary Experience

everything is connected, the future is not fixed. We are responsible for the consequences of our actions. What we do, matters!

This is true on a simple, practical level. As a responsible participant in God's ongoing creation, I work for a healthier planet and a healthier body. I can eat less and exercise more. The greatest epidemic in the United States is not Aids or hepatitis but obesity. I can walk instead of drive if it is a short distance or take my bicycle instead of the car. I can eat more fresh fruit and vegetable and less red meat. It is better for my body, better for my brothers and sisters and better for the planet. Transferring plant protein to animal protein wastes half of the available protein. My body benefits if I eat less meat. My eating less meat makes more available for others and assists in a more equitable distribution of our food resources. There is enough food for all to eat. I can share from my abundance with those who have less. I can work for a global community where all have enough. When we care for the poor as much as we care for our fashion, when we care for the planet as much as we care for our property, when we work for peace as much as we work for our possessions, we will have peace and prosperity throughout our planet. As the saying goes, "I cannot do everything, but I can do something. I will do what I can do!"

I was asked once why we had so much cancer in our society. I responded unexpectedly, saying it was because we chose to. They didn't understand, so I explained. We could find cures and treatments for our ills if we chose to invest our resources in that way. Our nation spends about as much on our military as the rest of the planet combined; the actual ratio varies from year to year. If we diverted just a portion of what we spend preparing to annihilate others toward finding ways to heal others, we would doubtless succeed in finding those cures. We have the problems we choose to have. We choose to suffer from cancer, hunger and other conditions in order to maintain a military machine that is overwhelming against any conceivable opposition rather than curing the world's ills and therefore perhaps needing a smaller military presence.

One of the Beatitudes is usually translated "Blessed are the meek, for they will inherit the earth" (Mt. 5:5). I like to note that the Greek word translated as "meek" in the Gospel is described by Aristotle in his virtue ethics as being neither too aggressive nor too passive but being perfectly angry: at the right time, for the right reason, with the right expression, toward the right goal, in the right manner, for the right cause. In every way, righteous anger; not an attitude we would normally describe as "meek"! I like to translate this verse as "Blessed are those who walk gently upon the earth; they will have an earth on which to walk."

Christ and the New Cosmology

The discoveries in astronomy and physics of the last century have radically altered and greatly expanded our sense of the universe within which we live, on both a macro and a micro scale. The universe is far older and far larger than we once thought. Created matter is much more complex and chaotic on a sub-atomic scale than we once imagined. It is a strange and weird place within which we live!

As our experience of the universe expands and our understanding of the universe changes, it is only logical that our understanding of God likewise expand. As Thomas Aquinas said, we cannot understand the Creator without first understanding creation. Creation, the universe, is a revelation of the divine before us. To study the universe is to explore the mind of God. As Einstein, a scientist, described his study, "I want to know God's thoughts. Everything else is details."

The new cosmology gives us a universe that is fifteen billion years old and therefore fifteen billion light years in expanse. It is a universe of probability and possibility rather than certainty, a universe that is intimately and inherently inter-related; everything is connected to everything and separation is an illusion. It is a universe in a constant state of flux. As Heraclitus declared 2500 years ago, "Everything changes, nothing remains." It is a universe of quantum physics, chaos theory and morphic fields.

This new understanding of the natural world gives us new language to describe all that we have experienced. It has given us a new understanding of physics, astronomy and biology. It has led us into far greater use of information technology. It has expanded our efforts in medicine in tests and treatment, CAT scans, genetic manipulation, etc. Margaret Wheatley has written in *Leadership and the New Science* how the new cosmology relates to new understanding and techniques in business and management. How might it relate to our theological expressions, particularly how might the new cosmology refresh our Christology? How can our new understanding of creation relate to our understanding of Christ?

There is much study and writing on the quest for the historical Jesus, some serious and some quite speculative. The quest for the historical Jesus is an interesting task, but also one that is limited. The historical Jesus lived two thousand years ago. His historicity is buried under twenty centuries of theology and teaching, doctrine and dogma. The man Jesus of Nazareth may never fully be known. The person of the Christ is one that may ever be renewed.

"Christ" is Greek for the Hebrew "Messiah," meaning "the Anointed One." In all the Old Testament there is only one person given this title: Cyrus the Mede. None of the kings of Israel or Judah. A Gentile and pagan, not a Jew. He is hailed as Anointed One because he allowed the Jews to return to their homeland after the Babylonian Captivity. It is interesting that the Jews give this sacred title to one who is not one of their own, but an outsider and a stranger to their faith.

The followers of Jesus give him this title because they have experienced the divine presence in a powerfully unique way in and through his presence. They experienced Jesus as more than a rabbi, a teacher, a prophet or a healer. He was all of those, yet also much more. So much more that they confessed the divine presence fully in him. Jesus of Nazareth was a historical person, a human being now lost to history. Jesus the Christ stands ever before us,

Creation in Contemporary Experience 57

as a divine sign and living symbol, as witness and wonder of the divine presence, purpose and promise before all human existence.

One concept of the New Cosmology that wonderfully expresses our understanding of Christ is that of Singularity. Singularity means something that is singular, unique. Certainly we believe that Christ is unique!

A Gravitational Singularity is a point in the space-time continuum where gravitational forces cause matter to have near infinite density and near zero volume: massively heavy and incredibly small. The Penrose-Hawking theory in general relativity sees gravitational singularities as creating black holes, which are in the center of every galaxy and around which the galaxy orbits. A singularity creates a complete restructuring of information. A temporal singularity is an event that creates a divergent narrative and may even lead to making a parallel universe.

The Big Bang is a perfect example of a singularity, a unique event that creates the entire universe. Black holes are another example of singularities, as each back hole attracts and builds the galaxies that orbit around them.

In human terms, Christ is also a singularity. Christ is a unique event that separates all human time into B.C. and A.D., before Christ and Anno Domini, in the year of our Lord. Now it is politically correct to say Common Era and Before Common Era, but it is Christ who has made the era to be common! If we go to Cairo, Egypt or New Delhi, India or Shangai, China or Tokyo, Japan and ask the person on the street what year it is, they will give us the year according to Christ. Christian and non-Christian around the world still count time according to Christ. The analogy of the Christ event as a singularity is not a perfect one, but it has had such a profound impact that it is certainly a singularity.

Christ spoke of a new era, a new time, called the Kingdom of God. This kingdom was not of this world, but a new realm breaking in upon this world. In brief, it is a world ruled by love: love God, love neighbor, love one another, love self, love enemy, love friend and stranger. It is a world that satisfies the great challenges of scripture: to care for the planet, to care for the poor and to work for peace. It is a world of shalom, where there is peace and plenty for all, a world where we share life's abundance rather than hoard it. Where it has been faithful, the church has been this New Creation, this Kingdom of God, this Ethic of Christ. Christ has begun a new era, a new world, a new way of keeping time. Christ has attracted all humanity in such a way that he has become the center of our history. No person has affected human history as this one has. Christ is our black hole, wherein all the world is a galaxy orbiting around this cosmic center.

Diarmuid O'Murchu in *Catching Up with Jesus* challenges us to do just that. This Christ is ever before us, ever beyond us, calling to us and challenging us to catch up and keep up. Brian Taylor in his twin volumes *Becoming Christ: Transformation through Contemplation* and *Becoming Human: the Core Teachings of Jesus* observes that as we grow into Christ and mature as Christians we also become more fully human in the process.

The Jesus of history may be buried beyond recovery, but the Christ of faith is ever alive, a singularity that has broken time itself and now breaks into our lives, making all things new!

A Contemporary Expression of the Eternal God

I was in a library doing research for my Doctor of Ministry project when I happened upon a book on quantum physics. I'd been curious about the topic for some time, so I checked the book out and read it. About a third of the contents were mathematical formulae that I skimmed, but I was excited with the words. I read more on the topic, and what I found intrigued me. I had read widely in the Christian mystics, and it seemed that the world of the mystics and the world of quantum physics was the same world. Different starting points, different perspectives, totally different vocabulary, but the same world being described. Others have found the same observation to be appropriate. The world of the great mystics of the ages and that of contemporary scientific theory is one and the same Creation.

How do the tremendous scientific breakthroughs of the past century, which have so profoundly reshaped our understanding of the universe and how it works, affected our understanding of God? The theory of evolution and the Morphic Field theory have changed how understanding of the biological world. Quantum physics has changed our understanding of the material world. Chaos theory has changed our understanding of causality. How do these changes relate to our theology, our understanding of God?

Einstein warns that religion without science is lame and science without religion is blind. Richard Dawkins' critical error in *The God Delusion* is that he attempted to freeze our understanding of God in some long abandoned stereotype. I am grateful for an upbringing that taught me that faith and reason are not mutually exclusive, that I can think and I can believe and those two disciplines support one another rather than oppose one another.

As our experiences of God and of God's world expand, so must our understanding of God and of God's world. As our experiences broaden, our understanding deepens. This is not to say that God changes, only that our understanding of God changes. Our understanding of God is always limited by the very fact of our finiteness. We are finite creatures attempting to describe the infinite Creator. We are three dimensional creatures, living in a perceived fourth dimension of time, attempting to describe that which has no dimension and supersedes all dimensions.

The question is "What do I mean by the word 'God'?" It might be good to begin by saying what I do not mean.

I do not mean some big guy in the sky, some cosmic superhero with a big "G" on his chest. I do not mean Zeus wielding lightning bolts from the sky or some storm creature fixing floods and causing quakes. I do not mean some tribal warrior picking sides in earthly struggles. God is not a being but beingness itself, God does not exist but God is existence, God is not a product of existence but the process that is existence. The problem with so many who deny or reject God is as J.B. Phillips described, *Your God is Too Small*. God is not some cosmic cop, some grand old man, some heavenly Santa Claus or magic genie in a bottle.

Any God that fits into my brain is already too small. Any God that I can define is inadequate. I do not set God's limits and determine God's nature. I can only describe my experience and offer my understanding, disclaiming from the beginning my inadequacy for the task.

If I were to describe the atom according to scientific understanding as late as the late 19th century, it would fail to meet the experiences and understandings we have gained in the past century.

Creation in Contemporary Experience 61

Would I then say atoms do not exist because my description of them does not match contemporary experience? Of course not! Nor would I take a medieval understanding of God, say it does not match contemporary experience and understanding, and therefore God does not exist. Foolishness! The logical, consistent choice would be to come to a contemporary description of God that matches contemporary experience and understanding. This expects religion and science to inform each other and recognized that we are both thinking creatures and believing creatures, that thinking and believing both fit into what makes us whole.

Any description of God that can be expressed in human terms is by definition limited and therefore inadequate. We are finite creatures, limited and mortal. God is that which is infinite, unlimited, and eternal. Such a concept will always be beyond us, yet we are compelled to describe what we have experienced and struggle to come to some understanding, however inadequate it might be.

There is a Hindu parable of four blind beggars who experience for the first time an elephant. Each describes their experience. "An elephant is like a muscular snake." "An elephant is like a tremendous tree trunk." "An elephant is like a large flapping leaf." "An elephant is like a large rock warming in the sun." Each described in turn the trunk, the leg, the ear and the body of the elephant. The experience of the entire elephant was beyond their grasp. They could argue about which description was right or they could accept that each description accurately described each experience and gain a greater understanding by accepting different expressions from diverse experiences, while allowing that all the descriptions combined might not fully describe the entire entity. And this is just an elephant!

I understand God as ultimate Truth, and I am convinced that any search for truth is ultimately a search for God. As Christians we say that Christ is the Way, the Truth and the Life. We confess that God is light and in God there is no darkness at all. Darkness may be described as moral darkness, the darkness of death, the absence of light, or it may be described as the darkness of ignorance. I believe that any search for truth, in any discipline and from any beginning point, if it is pursued honestly and thoroughly, will lead inevitably and inexorably to God, as ultimate Truth. We need neither resist nor fear any legitimate pursuit for truth; it can only lead us to God.

To the biologist I would say God is that which causes to be. To the mathematician I would quote Michael Dowd, God is that set which includes all other sets and is not itself contained by any other set. To the psychologist I would say God is described in our quest for meaning, our search for purpose in our lives. In geometric terms, God is a circle whose center is everywhere and whose circumference is no where. To the physicist I say that God is that time when all time is now and that place where all places are here. To the cosmologist I say that God is all that is, that was, that will be, in all times and all places, all that is physical, intellectual, emotional, spiritual, energetically.

The theory of conservation of energy in the universe, $E=mc^2$, says that beneath all matter there is energy. The confirmation of Higgs' boson, the "god particle," demonstrates how particles gain mass. In the ancient world, what we now call energy they would call spirit, a non-physical, immaterial force or influence. It should come as no surprise that modern science has proven the ancient mystics right: energy/spirit lies beneath, beyond and before all material things. George Lucas in *Star Wars* describing "the force" was not just writing fantasy but, using image and symbol, describing reality within reality.

To the linguist or the anthropologist, I would recall the Indo-European root behind our Germanic "god," and return to "gawatha," meaning "to cry out, to invoke," making the root to describe prayer, the calling out to God, more than God's own self and more a verb than a noun. Certainly not a name! "God" denotes what cannot be described, much less defined. Like moths to a flame, we drawn to what Rudolph Otto described as "the Mysterium Tremendum" in *The Idea of the Holy*, that presence before which we tremble. My favorite description of the divine comes from the 14th century classic of English mysticism, *The Cloud of Unknowing*, God is that cloud of absolute beingness, the more I enter into, the less I know.

Because words by their nature are inadequate, the language of God needs to be the language of poetry and symbol, language that Michael Dowd describes as "night language" rather than the "day language" of science. The language of science needs to be explicit. The language of faith needs to be expansive. The two are different, and that difference is necessary so that, working together, they can more fully describe the fullness of reality that we experience.

Though within inches of one another, my right eye and my left eye see from slightly different perspectives. That difference is essential for me to have any depth perception. The different perceptions of science and faith allow us to gain a greater sense of the depth of our human existence in this incredible, incredulous universe.

My favorite Greek philosopher is Heraclitus, who lived in the 4th century BCE in what is today Turkey. He wrote in aphorism, so that the reader is required to think about what he means and interpret what he says, joining in the philosophical task. I like to think that Heraclitus was the first process theologian.

Heraclitus wrote "You can never step twice into the same stream." By the time our second foot enters the stream, the water has already moved on, thus changing the stream. He wrote of the constancy of change, "Everything changes; nothing remains."

Jumping to the 20th century, Alfred North Whitehead used process as a way of describing the divine nature and the working of the divine will, the process itself is divine. The process of nature, of creation and evolution, of human history. God is at work in the world, both immanent and transcendent, both proactive and responsive. Hebrews declares "Jesus Christ, the same yesterday, today and forever." The Old Testament is filled with occasions when God changes, God turns, God even repents. God is very much "both-and."

In *The Evolution of God*, Robert Wright surveys Judaism, Christianity and Islam describing how our experiences of God, our understandings of God and our expressions of God have all changed, evolved, throughout history. It is not that God changes but that our understanding and expression of God changes as our experience of God changes.

Modern science informs contemporary theology. As we explore nature, we discover nature's God. Astrophysics, the Hubble telescope and the Big Bang theory give us an immense ever expanding universe that is fourteen billion years old and fourteen billion light years big. Particle physics gives us a universe that tiny beyond our understanding. Quantum physics gives us a universe that is dynamically energetic, incredibly interconnected and still largely unknown to us. Chaos theory sets us free from the prison of certainty and casts us adrift in sea of possibility and probability, where small choices may have large consequences. Morphic field theory stresses the interconnectedness of everything. Evolution reminds us everything is in a state of flow, the universe is increasingly complex and diverse and everything is changing.

All of this tells us of the nature of God. God is eternal, expansive, transcendent, immanent, ultimate, absolute, processing, responsive, more than we can experience, understand or express, more than we can think or imagine or hope. God is always that which is more!

Postscript

Congratulations on completing the text! I hope I have given you something to think about. If you are interested in more, I recommend the other volumes in this series: Harold Weiss for scripture study, Edward Vick for theology, Tony Mitchell for science, and Robert Cornwall for pastors looking for ways to incorporate a more contemporary world view into the worship experience.

I also recommend Judy Cannato's *Radical Amazement* and *Field of Compassion*. Judy writes in a way that truly invites the reader into the wonder and delight of God's creation. Michael Dowd's *Thank God for Evolution* opens the dialogue between faith and science in a way that is exciting and fresh. I am grateful for his insights. Brian Swimme is well noted in the new cosmology. John Polkinghorne is a theoretical physicist, theologian and Anglican priest who has written extensively of cosmology, physics and faith.

It is an exciting field and a great time to be alive. We are only beginning to comprehend how much we don't know, how vast and wondrous this holy creation truly is!

BIBLIOGRAPHY

Al-Khalili, Jim. *Quantum: A Guide for the Perplexed.* Weidenfeld and Nicolson. London, England. 2003.

Anonymous. *The Cloud of Unknowing.* Paulist Press. New York. 1981.

Beck, Richard. *The Authenticity of Faith: The Varieties and Illusions of Religious Experience.* Abilene Christian University Press. Abilene, KS. 2012.

Bruteau, Beatrice. *God's Ecstasy: The Creation of a Self-Creating World.* Crossroad Publishing, New York. 1997.

Cannato, Judy. *Field of Compassion: how the New Cosmology is Transforming Spiritual Life.* Sorin Books. Notre Dame, Indiana. 2010.

Radical Amazement: Contemplative lessons from Black Holes, Supernovas, and other Wonders of the Universe. Sorin Books. Notre Dame, Indiana. 2006.

Quantum Grace: Lenten Reflection on Creation and Connectedness. Ave Maria Press. Notre Dame, Indiana. 2003.

Collins, Francis. *The Language of God: A Scientist Presents Evidence for Belief.* Free Press. New York. 1996.

Cornwall, Robert. *Worshiping with Charles Darwin.* Energion Publications. Gonzales, Florida. 2013.

Darwin, Charles. *The Origin of Species: By Means of Natural Selection or the Preservation of Favoured Races in the Struggle for Life.* Grammercy Press. 1995.

Davies, PCW and JR Brown, editors. *The Ghost in the Atom.* Cambridge University Press, Cambridge, England. 1986.

Davis, Stephen. *God, Reason and Theistic Proofs.* Eerdmans. Grand Rapids, MI.1997.

Dawkins, Richard. *The God Delusion.* Mariner Books, New York. 2008.

de Chardin, Pierre Teilhard. *The Divine Milieu.* Harper Colophon Books. New York. English translation c. 1960.

_____. *The Phenomenon of Man.* Harper Classics. New York. 2008.

Delio, Ilia. *The Emergent Christ.* Orbis Books. Maryknoll, New York. 2001.

Dowd, Michael. *Earthspirit: A Handbook for Nurturing an Ecological Christianity.* Twenty-Third Publications. Mystic, Connecticut. 1992.

_____. *Thank God for Evolution! How the Marriage of Science and Religion will Transform your Life and our World.* Council Oak Books. San Francisco, California. 2007.

Edgar, Bob. *Middle Church: Reclaiming the Moral Values of the Faithful Majority from the Religious Right.* Simon and Schuster. 2006.

Ford, Kenneth W. *The Quantum World: Quantum Physics for Everyone.* Harvard University Press. Cambridge, Massachusetts. 2004.

Gellman, Jerome. *Experience of God and the Rationality of Theistic Belief.* Cornell University Press. Ithaca, NY. 1997.

Gleick, James. *Chaos.* Penguin Group. London. 1987.

Greene, Brian. *The Elegant Universe: Superstrings, Hidden Dimension and the Quest for the Ultimate Theory.* WW Norton and Company. New York, New York. 1999.

The Fabric of the Cosmos: space, time and the texture of reality. Alfred A. Knopf. New York, New York. 2004.

Hamer, Dean. *The God Gene: How Faith is Hardwired in our Genes.* Anchor Books. 2005.

Hawking, Stephen. *A Brief History of Time.* Bantom Books. New York. 1998.

_____. *The Universe in a Nutshell.* Bantam Books. New York, New York. 2001.

_____. *The Grand Design.* Bantam Books, New York. 2010.

Lewis, C.S. *Mere Christianity.* Fontana Books. London. 1960.

Lorenz, Edward N. *The Essence of Chaos.* University of Washington Press. Seattle, Washington. 1993.

Morris, Richard. *The Universe, the Eleventh Dimension, and Everything: What we know and How we know it.* Four Walls Eight Windows, New York. 1999.

O'Murchu, Diarmuid. *Quantum Theology: Spiritual Implications of the New Physics.* Crossroad Publishing. New York, New York. 2000.

_____. *Catching up with Jesus: A Gospel Story for Our Time.* Crossroad Publishing. New York, New York. 2005.

Otto, Rudolph. *The Idea of the Holy.* Oxford University Press. New York. 1965.

Phillips, J.B. *Your God is Too Small.* Simon and Schuster. New York. 1998.

Polkinghorne, John. *Belief in God in an Age of Science.* Yale University Press, New Haven, CN. 1998.

Rae, Alastair. *Quantum Physics: Illusion or Reality?* Cambridge University Press, Cambridge England. 1986.

Ridley, BK. *Time, Space and Things.* Cambridge University Press. Cambridge, England. 1976.

Sanguin, Bruce. *Darwin, Divinity, and the Dance of the Cosmos: An Ecological Christianity.* CopperHouse Books, Kelowna BC Canada. 2007.

Sardar, Ziauddin and Iwona Abrams. *Introducing Chaos.* Totem Books. Cambridge, England. 1998.

Stearns, Richard. *The Hole in our Gospel: What does God expect of us? The answer that changed my life and might just change the world.* Thomas Nelson. Nashville. 2010.

Steward, John. *Evolution's Arrow: The Direction of Evolution and the Future of Humanity.* Chapman Press, Canberra. 2000.

Swimme, Brian. *The Hidden Heart of the Cosmos: Humanity and the New Story.* Orbis Books. Maryknoll, New York. 1996.

_____. *The Universe is a Green Dragon: A Cosmic Creation Story.* Bear and Company, Rochester, Vermont. 1984.

Taylor, Brian. *Becoming Christ: Transformation through Contemplation.* Crowley Publications. Lanham, MD. 2002.

Becoming Human: Core Teachings of Jesus. Crowley Publications. Lanham, MD. 2005.

Vick, Edward W. H. *Creation: The Christian Doctrine.* Energion Publications, Gonzales, Florida. 2012.

Wallace, Mark. *Finding God in the Singing River: Christianity, Spirit, Nature.* Fortress Press, Minneapolis. 2005.

Weiss, Harold. *Creation in Scripture.* Energion Publications. Gonzales, Florida. 2012.

Wessels, Cletus. *The Holy Web: Church and the New Universe Story.* Orbis Books, Maryknoll, New York. 2000.

Wheatley, Margaret J. *Leadership and the New Science: Discovering Order in a Chaotic World.* Berret-Koehler Publishers, San Francisco. 1999.

Wilber, Ken. *Quantum Questions: mystical writings of the world's greatest physicists.* Shambala Publications. Boston, Massachusetts. 1985.

Wieman, Henry Nelson. *Religious Experience and Scientific Method.* Southern Illinois University Press. Carbondale, IL 1954.

Williams, Garnett P. *Chaos Theory Tamed.* Joseph Henry Press. Washington, DC. 1997.

Wright, Robert. *The Evolution of God.* Little Brown and Co. New York. 2009.

Also from Energion Publications

It is easy to imagine a congregation, gathered around a table covered with good fruit, asking each other the questions that arise from this study. Christ's presence is known in sharing such a meal. Come to the table!

Rev. Michael Mather
Broadway United Methodist Church,
Indianapolis, IN

This should be required reading for all clergy early in their careers.

Susan Nienaber
Senior Consultant
The Alban Institute

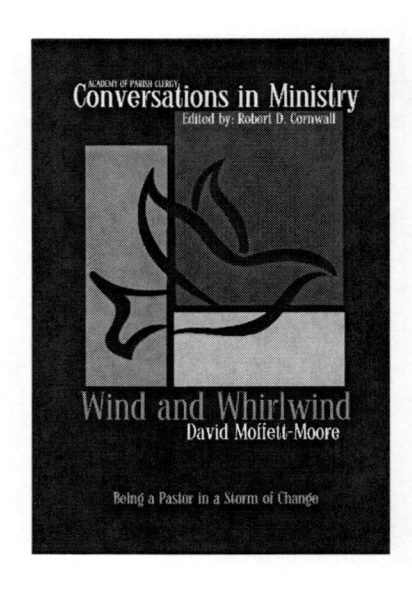

Questions about Faith and Science?

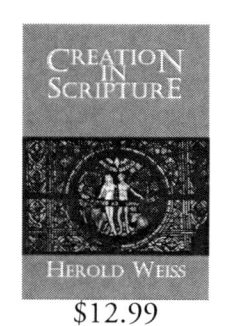

SCRIPTURE

Creation in Scripture — Herold Weiss — $12.99

THEOLOGY

Creation: The Christian Doctrine — Edward W. H. Vick — $12.99

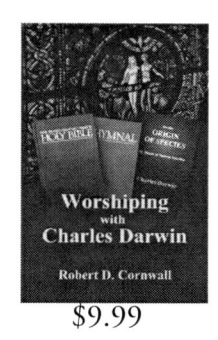

LITURGY

Worshiping with Charles Darwin — Robert D. Cornwall — $9.99

SCIENCE

Creation: The Science — $12.99
(Forthcoming, Fall 2014)

LIVING

$9.99

MORE FROM ENERGION PUBLICATIONS

Personal Study
Finding My Way in Christianity	Herold Weiss	$16.99
Holy Smoke! Unholy Fire	Bob McKibben	$14.99
The Jesus Paradigm	David Alan Black	$17.99
When People Speak for God	Henry Neufeld	$17.99
The Sacred Journey	Chris Surber	$11.99

Christian Living
Faith in the Public Square	Robert D. Cornwall	$16.99
Grief: Finding the Candle of Light	Jody Neufeld	$8.99
My Life Story	Becky Lynn Black	$14.99
Crossing the Street	Robert LaRochelle	$16.99
Life as Pilgrimage	David Moffett-Moore	14.99

Bible Study
Learning and Living Scripture	Lentz/Neufeld	$12.99
From Inspiration to Understanding	Edward W. H. Vick	$24.99
Philippians: A Participatory Study Guide	Bruce Epperly	$9.99
Ephesians: A Participatory Study Guide	Robert D. Cornwall	$9.99

Theology
Creation in Scripture	Herold Weiss	$12.99
Creation: the Christian Doctrine	Edward W. H. Vick	$12.99
The Politics of Witness	Allan R. Bevere	$9.99
Ultimate Allegiance	Robert D. Cornwall	$9.99
History and Christian Faith	Edward W. H. Vick	$9.99
The Journey to the Undiscovered Country	William Powell Tuck	$9.99
Process Theology	Bruce G. Epperly	$4.99

Ministry
Clergy Table Talk	Kent Ira Groff	$9.99
Out of This World	Darren McClellan	$24.99

Generous Quantity Discounts Available
Dealer Inquiries Welcome
Energion Publications — P.O. Box 841
Gonzalez, FL_ 32560
Website: http://energionpubs.com
Phone: (850) 525-3916

CPSIA information can be obtained at www.ICGtesting.com
Printed in the USA
LVOW12s0403110216

474499LV00001B/17/P

9 781631 990106